W9-ACN-218

PLANTS, MAN *and* LIFE

PLANTS, MAN *and* LIFE

by Edgar Anderson

WITH ILLUSTRATIONS

UNIVERSITY OF CALIFORNIA PRESS
BERKELEY, LOS ANGELES, LONDON
1971

ISBN: 0-520-00021-8
LIBRARY OF CONGRESS CATALOG CARD NO. 52-5870

UNIVERSITY OF CALIFORNIA PRESS
BERKELEY AND LOS ANGELES
CALIFORNIA

UNIVERSITY OF CALIFORNIA PRESS, LTD.
LONDON, ENGLAND

THIRD PRINTING, 1971

PRINTED IN THE UNITED STATES OF AMERICA

To

Oakes Ames

Orland White

Carl Sauer

who turned my attention in this direction

Preface

WHEN I STARTED TO WRITE THIS BOOK in the 1940s, I was given the good advice: "Don't write for an imaginary public. Think of some actual person as your reader; write the book for him." I knew the kind of man I wanted to interest. Ever since my late teens I had been explaining botany to visitors at various botanical gardens. Those I most enjoyed had deep-seated curiosity; good, disciplined minds; broad interests; but little technical understanding of plants.

Whom to choose as the perfect example? Pandit Nehru of India came to mind, so I kept him in my thoughts throughout the writing. It was years after the book appeared before I knew it really appealed to such readers. In stacks of fan mail the long, intelligent letters were from a dean of research in a medical school; the engineer of a transcontinental train; the quiet, astute wife of a leading scientist; the research department head of an international food-grain business; a telephone company executive; and so on.

It was *not,* however, the book my publishers set out to get. They had accurately detected a ground swell of interest in the story of the plants by which man lives; an interesting digest of what botany knows about the subject should have a ready sale. I presented them instead with a detailed exposition of what even the authorities did not know. Various simple facts which bore on the problem had not even been gathered! Important technical information of new kinds was piling up rapidly, but no one was scanning the whole wide field to see how everything might fit together.

At first the editors tried to keep me on the track; ironic overtones in the opening lines of chapter II indicate my reaction. Fortunately for my book, the firm went through a violent crisis having nothing to do with me personally. Far graver problems took nearly all my sponsors' attention. Eventually they were most cooperative in publishing *Plants, Man and Life,* the obverse of the book they originally planned.

Therefore, except for the four words they chose for a title, it all grew from my deep concern that here was a field needing talking about and looking into. There were few books on library shelves from which this story could be put together. For fellow scientists as much as for the general public, I drew on personal observations and hunches in building up an outline. European greens in autumnal landscapes (pp. 9–12), the tangled and continuing history of weeds and cultivated plants (pp. 13–15, 23–30), and the dooryard tangles in underdeveloped countries of fruit trees, fibre plants, vegetables, poisons, flowers, oilseeds, narcotics, vines and drugs (pp. 136–142) were unhackneyed examples often quite as unheard of by a professional botanist as by any other reader. By authorities in its special field the book was roundly praised or soundly damned. In some graduate laboratories it was read aloud after the senior professors had left. Those portions describing herbarium-minded botanists brought shouts of laughter. Among those who laughed was one of the young men whose cloud-forest behavior I had described (pp. 45–46). He told me about it the next time we met and we discussed what kinds of information are efficiently filed in an herbarium. It was give-and-take. We both saw more clearly the technical problems created by materials not ordinarily put on the same sheets as the specimens.

All this was 15 years ago. Enough botanists, ecologists, geographers, geneticists, and anthropologists who read the book as young men want their students to read it today, to justify a new edition. One or two of its basic ideas caught on and are now almost old-fashioned. Others of them are being developed at the moment among students who call themselves ethnobotanists. In another decade one of them (perhaps with one of his students) may bring out at last a balanced, accurate, completely revised edition. I hope so.

Edgar Anderson
February 1967

Contents

Illustrations

PLANTS, MAN *and* LIFE

I
Man and His Transported Landscapes

THIS HAS BEEN A WET SEASON; in the edge of our clover field in the flood plain there is more rib grass than red clover. Its narrow leaves, strongly creased at the heavy veins, stand up in stiff rosettes of dull green. Its slender-stalked flower heads, about the size and shape of the end joint on your little finger, are held daintily above the meadow. While they are in flower a circle of pale yellow stamens stands out from the flower head like the ornamental tuft at the back of a peacock's head. Aside from the functional grace of its general aspect, rib grass is a very plain little weed. Although you must have seen it in many lawns, although it has been intimately associated with man for a very long time, you may never have heard a name for it. Botanically it is a plantain, *Plantago lanceolata* to be precise, and it is now common in the pastures and meadows (and too frequently the parks and lawns as well) of Europe and America. Humble though it may be, it is a good plant for us to begin with because it has recently been found to be one of the most significant indicators of the long interrelationship of plants and man.

Iversen, a Danish botanist, is just now beginning to present his exact evidence for the first appearance of such meadow and field plants in northern Europe. These new facts come from a strange and laborious backhanded method of studying prehistoric climates, by examining pollen grain deposits under the microscope.

Pollen grains, the almost invisible source of your hay fever, are actually the carriers of the male sex cells of the higher plants. Some kinds of pollen, by fairly precise mechanisms, are transported to the waiting female plant by bees or hummingbirds. For many species, and this includes the bulk of our forest trees, the grains are merely released in such copious showers of golden dust that some few of them must inevitably reach the exposed female stigmas of nearly every plant. The bulk of this pollen shower is blown hither and yon, even into densely populated cities, where it makes miserable the lives of those unfortunate enough to be allergic to it. Even in a large metropolitan area the pollen grains of wild forest trees arrive in such numbers that the air is sampled by the Weather Bureau and daily pollen counts are published by the newspaper.

For this long flight through the drying atmosphere the biologically all-important cargo of the pollen grains is protected by a thin, glassy, plastic cover, a substance highly resistant to decay. It is this microscopic armor on pollen grains produced by the millions of millions which allows us to use them as an index of the past. As the wind blows them about, they fall as an invisible rain over the land. On lakes they settle gently to the ooze at the bottom. Some of them land on peat bogs where turfy mosses are growing upwards year by year, forming ever-thicker layers of peat. By boring down into peat bogs or into the beds of ancient lakes we can bring back to the laboratory narrow columns of ancient peat and muck, still adhering, layer by layer, in the sequence in which originally deposited. In amongst the soil are the ancient pollen grains which sifted down from the sky year by year, decade by decade, century by century. Washing out the pollen grains from the soil particles, identifying them under the microscope, and charting the numbers and proportions of the various kinds is laborious. It is one of those time-consuming routine chores which are the backbone of science. It gives a precise local record of vegetation going back into the glacial period, a record which can be cross-checked with geological data and with archaeological excavations. At the best sites good evidence from glacial terraces can be combined with the evidence from the pollen and with the position in a cultural sequence of the Bronze Age pottery and utensils which became buried in the peat.

It is rib grass which has given us the most insight into the relation between man and the vegetation of those ancient times. The pollen of rib grass is abundant and distinctive. Iversen finds there was none in Danish peat until the first agriculturists arrived in early Neolithic times. At that point there is sometimes a narrow band of charcoal in the peat, showing that the land must have been cleared by fire. Just above the charcoal, pollen of such woodland plants as ivy disappears completely for a time, whereas rib grass, previously unknown, becomes abundant and increases progressively, layer by layer. The sequence of peat with ivy pollen, covered with a layer of charcoal, and above that peat with rib-grass pollen is evidence for the succession of dense forests cleared by fire, followed by fields in which rib grass became increasingly abundant.

The whole spectrum of the pollen deposits is altered when these first farmers appear. Cereal pollens increase. Plants of the oak woodland lessen or disappear; birch pollen increases rapidly — it is one of the trees which can come in after an extensive burn. For the pollen record, the effect of early agriculture is as severe as a shift in the climate. In this earliest wave of farmer-folk the action was local but intense. It was much the kind of burn-over, plant, move-on agriculture which one finds so commonly in the mountains of Central and South America and eastern Asia. The Neolithic farmers moved on ahead from a clearing they were about to abandon, cleared the new site with fire, stayed for a few years, and then moved on to another clearing.

These early farmers kept to the richer oak woodlands. Much later in the Iron Age, a more extensive type of forest clearing took place which affected the sandy pinelands as well as the richer sites. The pollen record shows even greater changes in the landscape. The percentage of nonwoodland pollens rises higher and higher. When the sandy pinelands were brought into cultivation they were soon abandoned. Heather spread into the deserted sandy fields, and the extensive heaths, which still characterize the sandier parts of northern Europe, came into being.

Century by century and millennium by millennium, the pollen column records these changes in the forest cover. Early man introduced, intentionally, a few crop plants to northern Europe. Unintentionally he brought in alien weeds and changed the whole landscape with his burning and cultivating.

Where had the rib grass come from when it suddenly turned up in Denmark along with the Neolithic farmers? To that question we have little more of an answer than the modern farmer who merely knows that in spite of his best efforts rib grass turns up persistently in new fields and old. By one means and another it has managed to travel with man from country to country and even from continent to continent with no encouragement from him. For yarrow and some other meadow plants we have slightly better evidence. Another Scandinavian botanist, Turesson, traveled over Europe and western Asia, digging up samples of some of these common meadow plants and sending them back to his experimental plot in Sweden. One of his discoveries was that even for those species native to the mountains and forests of western Europe the strains of them now growing in European pastures and meadows may not have spread in from close at hand. Grown in Turesson's garden the meadow plants of Europe more closely

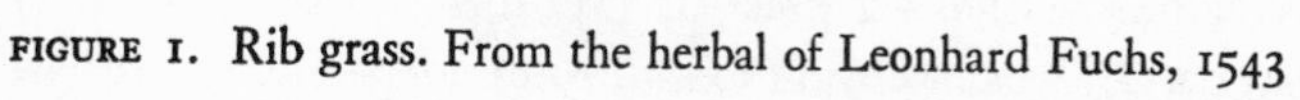

FIGURE I. Rib grass. From the herbal of Leonhard Fuchs, 1543

resembled the varieties of these same plants which are native far to the east in such places as the Altai Mountains than they did the transplants from European mountains. The meadow strains of yarrow, buttercup, and daisies may be older than European culture. In part at least they must have spread into Europe when the earliest cattle raisers began to press out of central Asia. They will be as exact evidence for European prehistory as long barrows and bell beakers, when our information about them is as precise.

Few Americans realize how completely our American meadow plants came along with us from the Old World. In our June meadows, timothy, redtop, and bluegrass, Old World grasses all three, are starred with Old World daisies, yarrow, buttercup and hawkweeds. The clovers too, alsike and red and Dutch, all came from the Old World. Only the black-eyed Susans are indigenous. An informed botanist viewing such a June meadow may sometimes find it hard to point out a single species of plant which grew here in pre-Columbian times.

Few of these aliens were deliberately introduced; one or two being noxious, we strongly resisted. The clovers were, in part, imported on purpose, and a few of the grasses, but the daisies, the buttercups, the yarrows, and the pasture thistles just moved in by themselves. Botanists frequently refer to these fellow immigrants as European, for their ancestors, like ours, moved to the New World from Europe. Yet our ancestors, biologically speaking, were not indigenous Europeans. They were mostly Asiatics who had been in Europe for what in biological terms had been a comparatively short time. How many of our dooryard plants trace back to Asia, which ones joined up with us in Europe, what few of them are authentic Americans? Most of these questions we cannot yet answer precisely. Yet these questions are only the first part of what we want to know in understanding the relationship of man and plants. In the lands where these fellow travelers joined our entourage, what led them to join up with us in the first place?

In many of our western states one drives for hour after hour and sometimes for day after day between long lines of wild sunflowers which, all untended, border the highways. In Idaho, in Wyoming, in the Dakotas, in Kansas and Nebraska, one frequently sees this double line of golden yellow leading ahead all the way to the distant horizon. Yet the grasslands on either side of the highway may be virtually free of sunflowers. In a dozen western states the highway winds through a grassy or shrubby landscape between parallel rows of sunflowers. What is there about man which makes him unconsciously adopt such plants as the sunflower? What is there about sunflowers which permits them to succeed along highways or in railroad yards or on dump heaps, but keeps them away from many native grasslands?

It is not until one sits down to work out precise answers to such questions that he realizes that unconsciously as well as deliberately man carries whole floras about the globe with him, that he now lives surrounded by transported landscapes, that our commonest everyday plants have been transformed by their long associations with us so that many roadside and dooryard plants are artifacts. An artifact, by definition, is something produced by man, something which we would not have had if man had not come into being. That is what many of our weeds and crops really are. Though man did not wittingly produce all of them, some are as much dependent upon him, as much a result of his cultures, as a temple or a vase or an automobile.

As a humble illustration of the transported landscapes in which we live, take the autumnal aspect of the central and eastern United States, roughly the region from Boston and Philadelphia to Minneapolis and Kansas City. In all that area, green in the autumnal landscape is a measure of European influence. The green grasses of pasture and roadside, the green trees of orchards and parks, are greens which have come with us from Europe. Our own native flora was bred for our violent American climate. It goes into the winter condition with a bang. The leaves wither rapidly, they drop off in a short time, frost or no frost. In their hurry many of them leave enough chemicals behind to give us brilliant fall color. Virtually all the autumnal green which one sees in the eastern and central United States is European. European trees and grasses color slightly and slowly if at all; our native grasses are as bright with color as our native trees. Perhaps I had better describe a few of the grasses in detail; for some reason grasses are so taken for granted (even by quite knowledgeable people) that few naturalists realize that European grasses (like European trees) stay green in the fall, with us as they did in Europe.

It is early autumn as I begin to write. The Ozark hills brighten day by day; the sumacs and Virginia creepers are in full color, the ashes are beginning to turn. Almost unnoticed the prairie grasses on the ridgetops and along the railroad are passing into their own exuberant display of fall color. In strong clumps the wiry stems of turkeyfoot rise shoulder-high. Their short stubby tassels branch out abruptly from the tips of the stems. Seen against the sky they really do look like a bird foot held up into the air. All summer these grasses have been a fairly uniform green. Now the narrow leaves are reddening, the stem is turning gray-blue, the tassel itself reddens into a strange bright brown like a Harris tweed. The little bluestem, a close relative and ancient prairie companion of the turkeyfoot, now shows why it too deserves the name the pioneer gave it. The stems and leaves are so richly patterned with color that one needs to take them home and view them against a neutral background fully to appreciate their beauty and intricacy. The stem and much of the leaves are overlaid with a bloom of celadon blue. At the joints, along the leaves, and in the silky tassel, are lines and splashes of Chinese red, rich brown, and bright yellow-green.

Nor are these two tall prairie grasses exceptional, unless it be in their commanding height. Our characteristic native grasses, like our characteristic native trees, color brilliantly and with dramatic suddenness before lapsing into winter dormancy. Like our trees they were bred in this violent continental climate with its cold clear winters, its heavy humid heat in early summer, and its autumnal frosts which set in so rapidly after the warmth of August. A late October cold wave does not catch them unprepared. They are physiologically attuned to shifting rapidly from the lush growth period of a violently hot summer to the complete dormancy required by a vigorous winter.

Drive through the countryside at this time of the year and note the distribution of the greens. The swamps and woodlands are all in full color, it is native American autumnal brilliance. It is the pastures, the strips of sod along the highway, the lawns, the orchards, the gardens, which are green. Their green comes from species which man has planted, like the apples and the lawn grasses, or from species which have been bred to withstand his coarse ways and grow untended along the highways and the fencerows. Note the contrast at those points where a highway carries one quickly from a woodland into a town. At the very edge of the town the greens begin to widen out and take a larger share in the landscape. Our cultivated apples, European trees, still hold their dark-green leaves. The English beeches in the parks ripen slowly as if they had nothing worse ahead of them than the cool mists and slow onset of an English winter. The English oaks are dull dark green. The Norway maples show no sign of autumnal color. The lawn grasses are green, greener than most of them have been all summer. They should be. They are north

European grasses. They too were bred for that slow cool creeping autumn which no American can understand until he spends that season in England. It is, in northern Europe, an autumnal change so gradual that only towards the Christmas holidays when an occasional spring jasmine or winter iris comes into bloom in a sheltered corner, does an American visitor realize that the fall is finally over and that winter is far enough along so that the earliest spring flowers are blooming. Our common bluegrasses are from Europe, even those kinds which are now so much at home that we call them Kentucky bluegrass and Canadian bluegrass. Our creeping bents are European. So are our fescues, our redtops, as well as our common dandelions, most of our plantains, and many another roadside weed. As autumn comes on and virtually all the native flora prepares for an American winter by hurrying into dormancy, the fresh greens which stay on in our landscapes are European immigrants like us. Some of them, such as the boxwood and the graveyard myrtles, we deliberately imported; most of them just came along. If you want to know how much of the landscape in which you spend your days is authentically American, look around your home town in midautumn. As man moves about the earth, consciously and unconsciously he takes his own landscape with him.

A transported flora which is quite as derived and perhaps even more ancient covers the rolling hills of coastal California. They are draped with a distinctive grassland which differs in make-up and in origin from our eastern meadows. Almost none of it was introduced deliberately; it, too, has followed man around the world, but by a different route. These coastal grasslands need to be seen for an entire season to be appreciated to the full. Their shape is distinctive; voluptuous curves contrast with rocky outcrops and occasional sharp summits. Some are nothing but grassland as far as the eye can see, others are set with crooked live oaks or lined along furrowing canyons with the evergreen leaves of the California laurel. When finally the winter rains start in earnest the grasslands come to life. They freshen into yellow green, as bright as new cheesecloth bunting. They grow up into a green plush of grass and weeds and wild flowers which ripen as the rains stop and the dry summer weather begins. One of the loveliest moments in the California season is when the pervading spring green is just being succeeded by the yellow of summer and fall. The gold spreads along the hilltops first, where the soil is driest. For a week or so the hills are parti-colored, golden along the ridges and outjutting flanks, fading into fresh green at the bottom of the slopes. A few more days and the rolling hills are a yellow brown, a shining golden yellow which catches the light and for eight months is a bright foil to the dull black-green live oaks.

The bulk of the plants in these coastal grasslands are not originally Californian. Many of them may have been there since before the days of the Forty-Niners, but they trace back to another part of the world with a similar climate and a much older civilization. They are Mediterranean weeds and grasses that started moving in with the earliest Spaniards and swept over the landscape, at times almost obliterating the original vegetation. The native grasses still persist here and there; most of the beautiful wild flowers are native but the bulk of the vegetational mantle is a gift, or a curse, perhaps both a gift and a curse, from the ancient civilizations around the Mediterranean sea. The plants which are growing unasked and unwanted on the edge of Santa Barbara are the same kinds of plants the Greeks walked through when they laid siege to Troy. Many of the weeds which spring up untended in the wastelands where movie sets are stored are the weeds which cover the ruins of Carthage and which American soldiers camped in and fought in during the North African campaign.

How did these Mediterranean weeds get to California in such numbers? We have little exact information but it is not hard to make a reasonable guess. As soon as livestock were brought in, the weeds traveled in the hay and in the seeds of field crops. Probably the introduction began with the very earliest Spanish visitors. When the sailing ships were loaded in the Old World their supplies would have been stacked up on the quay. Every time this was done a few little pieces of mud could have become caked on kegs and boxes or caught in the cracks. Most weed seeds are small. Hundreds of them could have traveled in every shipload. Of these hundreds a few lodged in the proper sort of spot when the ship was unloaded. California's climate is very similar to that of Spain, and in those days there would have been few native plants fitted to survive in the strange scars man makes on the face of the earth. The weeds brought in by the Spaniards already had much experience of man. Some of them had evolved through a whole series of civilizations, spreading along with man from the valley of the Indus to Mesopotamia and on to Egypt and Greece and Rome. Some had long histories behind them before they ever reached Spain, and for hundreds of generations had been selected to fit in with man's idiosyncrasies.

Walk to the outskirts of a coastal town in California and examine the flora item by item. The vacant lots in springtime are lined with the feathery plumes of wild fennel, an anise-scented herb something like a giant carrot. We do not know just when fennel began to travel around with ancient man, but it was long ago. Here and there are wild radishes, innocent of any useful root, but bearing flowers and seed pods closely similar to those one might find in a vegetable plot of radishes run to seed. No one can yet say exactly when and where radishes — weed radishes, vegetable-garden radishes — first became our traveling companions. Farther out of town the grasses outnumber the weeds and among the grasses the wild oats, of more than one species, outnumber everything else put together. When mother nature started sowing her ancient Mediterranean wild oats over coastal California she set the landscape pattern for years to come. It is the wild oats which grow so quickly after the rains have come. It is the wild oats which ripen so rapidly. Their graceful open tassels shake in the wind, reflecting the light by tiny pinpoints from their shiny glumes. It is these little oat glumes (the chaff around the grains) which, by the millions of millions, make the coastal hills glisten in the sun, a glowing golden plush. It is the wild oats which give the coastal breezes their distinctive sound if one can ever, in coastal California, get far enough away from his fellow man to hear nature's own noises for a few moments. The tiny metallic glumes clap against each other in the wind so that the hills give a rustle as of stiff silk petticoats when the breezes pass over them. It is these delicate and graceful grasses, ripened to tinder, which are one of the chief California fire hazards. In dry profusion they line the roadways, ready to flash into flame if a careless motorist forgets and tosses a glowing cigarette butt across the fence.

Fennel, radish, wild oat, all of these plants are Mediterraneans. In those countries they mostly grow pretty much as they do in California, at the edges of towns, on modern dumps and ancient ruins, around Greek temples and in the barbed-wire enclosures of concentration camps. Where did they come from? They have been with man too long for any quick answer. They were old when Troy was new. Some of them are certainly Asiatic, some African, many of them are mongrels in the strictest technical sense. Theirs is a long and complicated story, a story just now beginning to be unraveled but about which we already know enough to state, without fear of successful contradiction, that the history of weeds is the history of man.

II

The History of Weeds—A Detective Story

AT THIS POINT the intelligent reader, having been told something of the tangled history of weeds and of man, will be ready for an orderly unrolling of their story. Most likely he will expect to begin with the weeds of the Stone Age and then to pass on neatly to Iron Age weeds, Egyptian weeds, classical weeds, Holy Roman weeds, Elizabethan weeds, winding up at modern weeds, perhaps with a few final paragraphs about post-atomic weeds. Unfortunately this is not that kind of a problem, nor am I that kind of an author. Though all of what we shall be talking about is part of one piece, though it all belongs together, it is a story just naturally full of footnotes and asides. It is no job for an encyclopedist: for some parts of it the facts we need are not yet brought together; for others they have been assembled uncritically. The reader should be warned that as yet there is no one to whom he can turn for an orderly history of weeds; by a strange paradox these commonest of plants are comparatively unknown. My own special competence in this field is merely that of appreciating the kind of facts we must get before we can understand weeds and set ourselves to working out their histories. So if it is unfortunately true that I do not consider myself an authority on weeds, it is equally true that I do consider myself a real authority on what is not known about them. This book, in other words, will have to be much more like a detective story than a textbook. To be more exact, it will be a set of related detective stories in which suggestive pieces of evidence are brought in and analyzed but in which the final chapter finds us still running down likely clues, far enough along in our quest to describe the villain but not as yet able to point him out by name.

The vegetation of many a remote mountain range is better understood than the common flowers and weeds in your garden. Go climbing in the Rocky Mountains and bring back a collection of rare little alpines from the Continental Divide. There will be a score of experts who can name them for you. If you are naturally suspicious and bother to send your collection to three different experts, you will be encouraged at getting back practically the same set of names from all three. Bring back a collection from the Aleutian Islands or the mountains of Guatemala and you will fare nearly as well. If your holidays take you farther and farther afield, to the South Seas or the Congo, you may still get substantially all your collections named so long as you do not stray into a dooryard, a flower garden, a greenhouse or a dump heap. Make collections from such places and some intelligent botanists will refuse to put names on many of your specimens or will tell you that the plants in question are pantropic weeds and that names do not mean very much. Suppose, for instance, you pick one of the modern bearded irises from your own flower garden and send it to an experiment station or a university botany department with a request for the scientific name. Two out of three will probably tell you that your plant belongs to *Iris germanica,* which it most certainly does not. German flags, to use the name for *Iris germanica* which your grandmother used, are a group of earlier-flowering, old-fashioned varieties, most of which have been around for at least some hundreds of years, and Linnaeus gave them this scientific name in the eighteenth century. Our modern bearded irises are not directly related to them and they resemble

Iris germanica only superficially, but as yet no botanist has taken enough interest in their technical classification to provide them with a botanical name. If you have the good luck to send your specimen to an expert who is both honest and informed, he will tell you that your specimen is a tall-bearded iris, of hybrid ancestry, and without any recognized Latin name. The reasons for this paradox that the commonest plants are the least known makes a complicated story but one which I want to tell in a good deal of detail. But before we can get down to that subject understandingly, there are several simpler matters which must be taken up first. Probably the best of these to begin with is the story of the spiderworts, a good example of what happens to plants when man grows them in his gardens.

The spiderworts (Tradescantia to the initiated) have been known to botanists ever since American plants were first taken back to the scholars of Europe. You probably have seen spiderworts many times though you may not know them by name. They are an old-fashioned garden flower, rather on the weedy side. Their bright, three-petaled flowers (a somewhat electric shade of blue is the commonest color) are borne in small clusters. Though showy in the early morning, they distress gardeners by closing at midday, except in the dullest weather, so that all the afternoon and evening they are a rather untidy mess of twisting leaves, unopened buds and withered flowers. After they close, the flowers have a curious habit of turning liquid within the enclosing sepals. I was shown all this when I was a child and we used to make yesterday's flowers "weep" by pressing them with thumb and forefinger. When this was done, out would run one or two drops of deeply pigmented sap, a kind of purple ink quite capable of staining one's fingers or one's clothes. When there is a heavy dew the pigment colors the dewdrops which hang from the closed-up blossoms in the early morning. "Widow's-tears," one of the old-fashioned names for the plant, referred to these droplets, supposedly so named because they dry in a day. In European gardens, to European botanists, and in the botanical literature of the world, by far the commonest species of spiderwort would seem to be *Tradescantia virginiana.* When I came to study this group of plants critically with my colleague, Dr. R. E. Woodson, we found that what commonly passed for *Tradescantia virginiana* in Europe was by no means identical with the original species native to this country. Gradually we worked out its history, a simple one but worth telling because it is one of the few examples in which we know precisely the relationship between a cultivated plant and its wild progenitors.

Tradescantias were taken to England in Colonial times. In Parkinson's charming account of the garden plants of seventeenth-century England, published under the punning title of *Paradisi in sole,* we find their introduction credited to John Tradescant, head gardener to Charles the First. Parkinson described them by a long Latin phrase which he roughly translated as "the soon-fading Spider-wort of Virginia, or Tradescant, his Spider-Wort." Parkinson went on to say that the species was "of late knowledge, and for it the Christian world is indebted unto that painfull and industrious searcher, and lover of all nature's varieties, John Tradescant, who first received it of a friend, that brought it out of Virginia." In England spiderworts became favorite plants in cottage gardens but certain curious properties gave them a truly scientific career. Since the development of the microscope no other kind of plant has been so closely associated with botanical work over so long a time. Scientific interest began because of the delicate blue hairs which veil the stamens in a sort of ostrich-plume mist and which distinguish Tradescantia from most other plants. Under a microscope each of these tiny hairs becomes a chain of elegant blue beads, each bead a single cell, all of them so transparent that even in the living condition without any technical hocus-pocus, other than getting them in place under the microscope, one can see all the parts of the cell. Later studies showed that other cells in the plant, most particularly the germ cells, were also large and clear and particularly suitable for study under the microscope. Scientific reliance upon Tradescantia as research and demonstration material has continued unabated by reason of these remarkable qualities. Halfway through the twentieth century, we still find them being generally used as demonstration material for the education of premedical students and as subjects for determining the effects of various radiations upon living matter in such laboratories as those at Oak Ridge, Tennessee.

These were the circumstances which led me, some twenty years ago, to make a special study, not only of *Tradescantia virginiana,* but of all the species closely related to it. Its germ cells were being intensively studied in laboratories in Europe and America. Some of these researches were narrow and pertained only to the minutiae of cell organization. Others were much more ambitious and sought to establish general principles of evolution from a direct study of the germ plasm, since it is the physical basis of inheritance. In the nineteen-twenties brilliant inferences from observations on the cells of Tradescantia had produced a series of conflicting theories as to what the course of evolution might or might not be in that genus. It seemed to me that, fascinating though these theories were, some facts about the kinds of spiderworts and their distribution would be a fundamental background for all this discussion and I accordingly set out to monograph this group of species. Though I had had no training in this kind of work, I knew I could count on my colleagues because of the central importance of Tradescantia to so many biological problems, nor was I disappointed. Scores of them sent me living plants as well as pressed specimens, and my former student, Dr. Robert Woodson, undertook the lion's share of the tedious technical bibliographical chores which are the skeleton of a modern monograph—keys, descriptions, citations of specimens and the orderly presentation of references to previous work.

It was no ordinary monograph. We not only studied all the collections of pressed specimens which were available in the United States, we journeyed to the places where Tradescantias grow and studied and collected them in the field. Tradescantia is fortunately the kind of a plant that not even a botanist can kill and it was a simple matter to keep a living collection of specimens from our own field trips as well as those which were sent in by other botanists. At the same time, with my colleague, Dr. Karl Sax, an authority on the study of the germ plasm under the microscope, I started a survey of the most pertinent and easily ascertained facts concerning the germ plasms of these same plants. Nearly ten years were devoted to this work and so many thousands of miles of travel that I have never undertaken to calculate the total. It carried us to every state east of the Mississippi River as well as several west of it and it made us, as a group, more familiar with more varied aspects of evolution in one small group of plants than anyone had been before that time.

The general public have such odd ideas of how a scientist spends his time that an actual chronicle of one of these collecting trips might not be out of place. We traveled usually by motor, and if possible by station wagon, station wagons in those days being functional vehicles which served as carry-alls, field laboratories, and lodging for the night. Mostly we went by twos, sometimes by threes or fours, and though there were many short trips, the bulk of our information came from journeys of a thousand miles or so which took ten days or longer. A specimen day would start out early in the morning; spiderworts flower only before midday and he who is devoted to their study must not linger in bed. We would drive down the highway at a good clip, counting on our knowledge of the kind of place where Tradescantias were to be found plus the brilliance of their flowers to keep us from passing many which were within easy distance of the road. Sometimes we would drive a few miles, sometimes half a day, before finding the next patch. Then at the cry of "Trads" the auto would pull off beside the road and we would tumble out to make a quick reconnoiter of the site before setting down to our regular routine. One of us would dig up one or two plants to send back home to the experimental plot, put specimens into the plant press to dry as a matter of record, and take notes about flower color, the technical details of the plants, and a short description of the place in which they were growing. The other member of the party, usually me, worked with the microscope. Here was the special advantage of Tradescantia. To study the germ cells of most plants and animals takes days, if not weeks, of technical preparations, most of which require something like a laboratory. Tradescantia, for a variety of technical reasons, is so much more effective for microscopical analysis that with a field microscope, a few glass slides, a little bottle of stain and a pair of dissecting needles I could set up my elementary laboratory on a stump in a pinewoods or the cement culvert of a railroad track.

In pleasant weather it was a most delightful occupation. One had the stimulus of being out in country which was new to him and the added stimulus of exploring under the microscope as well as over the face of the earth. The cells which we studied were beautiful in themselves. The critical information could be gained either from the very young pollen grains or from the cells which produce them, the so-called pollen mother cells. Both of these kinds of cells were dissected from the anthers of tiny unopened buds just below the flowers. The pollen mother cells, when they could be located at the right stage, looked under the microscope like great crystal globes turning over and over in the red solution on the glass slide. Within them and darkening rapidly with the stain were the chromosomes, the fundamental unit of which the germ plasm is made up. One of our main concerns was to find whether the chromosomes were in sets of six or in sets of twelve and if they were behaving in an orderly way or were exhibiting any one of various aberrations which had been discovered in the world's botanical laboratories. So incredibly simple is the technical study of the germ plasm in Tradescantia that with average luck we could locate a colony to study, make all necessary notes, determine the chromosome numbers, pack everything back in the car and be on our way again all in an hour, to repeat the process in another woodland or along another cliffside, or on another mountaintop, a score or so of miles down the road.

One of the most important consequences of this enterprise I did not fully appreciate at the time I was doing the work. It gave me a unique insight into the problem of cultivated plants and weeds. Contrast my position with the student of wheat or of rice. Whence came these major crops? They have been grown since before the dawn of history. There are some weeds obviously related to wheat and to rice but some of these we already know to have been produced by the crop rather than being the weed from which the crop originated. Slowly and painfully the world's authorities on wheat and on rice are working into this problem, but it will be a long time before one can designate with certainty the exact wild progenitors from which they sprang and then set himself to the further task of determining to what extent they have been modified by their long association with man and by what processes this went on. Compared to these world problems, the evidence presented by my little spiderworts seemed almost too trivial a problem to take up in scientific circles. Spiderworts were simple garden flowers of ordinary folk, not even an important flower crop like snapdragons or roses. Here and there they persisted, uncared for, hardly bothersome enough to be called a weed and yet with enough weedy attributes to be put in that category if one wanted to split hairs. Yet they presented a unique opportunity to investigate this important problem of what happens to plants when they are domesticated. *Tradescantia virginiana* of European gardens had been in cultivation a bare three hundred years, yet it was no longer like the *Tradescantia virginiana* which grows wild from Pennsylvania to Missouri. Exactly what were the differences and how had they come about? For the first time in biological history, a student of a domesticated plant (albeit a most unimportant one) was thoroughly acquainted with that plant as it existed in cultivation and also *with all its possible wild progenitors.*

As I continued to become more and more familiar with these two *Tradescantia virginianas,* the cultivated strains and the wild ones from the eastern United States, it became more and more apparent that they were two different sorts of things. Some of these differences were highly technical: the cultivated plants had a high frequency of irregularities and aberrations in the organization of their germ plasm, the wild ones had practically none. Many of the differences, however, were simple everyday matters that any unlettered gardener might have noticed. The wild plants of *Tradescantia virginiana,* even after they had been cultivated for several years in my experimental gardens, still behaved like most spring wild flowers. They rushed into bloom quickly, went out of bloom quickly, and then the whole top of the plant yellowed off and died down in midsummer. The cultivated ones did nothing of the sort. They did not rush into flower; they continued in bloom well into the summer and if they were well taken care of, the foliage persisted until the frost cut it down in late autumn.

Just as I was beginning to ponder over such differences as these, some of the things which happened in my experimental garden gave me the key which explained much of what had been happening to *Tradescantia virginiana* during its three-hundred-year career as a garden plant. The experimental garden was made up of several plots and one of them was given over to two species of

FIGURE 2. One of the various American spiderworts which entered into the ancestry of the so-called *Tradescantia virginiana* of our gardens.

Tradescantia in which we were particularly interested. One of these looked rather like *Tradescantia virginiana.* It was then known as *Tradescantia reflexa,* though scholars have since determined that the name *Tradescantia ohiensis* is to be preferred because of its priority. This ubiquitous plant of the Middle West has no common name, though "railroad-track spiderwort" would be an appropriate one. It is the lusty plant with a waxy bloom over its leaves whose blue or purple blossoms brighten morning roadsides and railroad tracks in the early summer. The other species, *Tradescantia subaspera* var. *pilosa,* commonly known in the horticultural literature as *T. pilosa,* is so different from our other spiderworts that the first time I found it, I did not know that it was a spiderwort. In the first place it was growing in deep, shady woods with rich soil. In the second place, it had broad hairy leaves and looked as if it might be some peculiar kind of lily. Finally, though it was already early summer, its flowers were not yet open. When they did come out a fortnight later, I recognized them as Tradescantia flowers, though they were only half the size of those on the other spiderworts. These two species, *pilosa* and *ohiensis,* occur near each other, the one in deep woodland, the other on rocky places and along railroads, over thousands of square miles in the Middle West. One of the technical questions we had been interested in was why they so seldom crossed with one another even when growing close together. Hybrids were so rare that in days of looking we found only one or two plants and those only in places where the woodlands had been cut over and heavily grazed.

In the experimental garden the story was very different. Though *pilosa* is normally a woodland plant, it thrives in a perennial border and in my little plot it would be hard to say which species was the happier, *pilosa,* or the railroad track spiderwort. Both of them blossomed amazingly and went to seed in abundance. Within a few years, seedlings started coming up throughout the plot and became quite a problem. We let some of them develop to see what they would turn out to be, and we raised a whole batch experimentally for a double check. To our surprise, many of them turned out to be hybrids. In their general appearance the hybrids were intermediate, on the whole, between the two parent species. Their leaves were neither as long as those of *reflexa* nor as broad as those of *pilosa.* The plants did not bloom as early in the season as *reflexa* nor as late as *pilosa.* They were not cabbage-leaf smooth like *reflexa* nor thickly felted with hairs like *pilosa* but bore scattered hairs on the stems and leaves and many short hairs on the outside of the blossoms.

Now that we knew what to expect when the two species were crossed we went back to some of the places in which we had found both species growing nearly as close together as in my experimental plot and made a second careful look for hybrids in the wild. In most areas none were to be found. This was most illuminating. There were no hybrids in the wild, not because the two species were not cross-fertile, nor because bees did not carry pollen back and forth, but because in the strict interlocked economy of nature there was no room for something different. There was no niche into which they could fit. The only hybrids we did come upon were in those places where lumbering and pasturing had so destroyed the natural balance of things that there was an opportunity for something different to get started. In the experimental plot, carefully weeded spring and fall, the hybrid seedlings were at no disadvantage. They had no perfectly adjusted native plants to compete with them when they started to germinate and grow. A plant is up against a new kind of evolution once you move it into a garden. It still has a battle to fight but in a different way. So long as it produces variants which man finds pleasing and amusing, it may survive, though it be less lusty, less capable of caring for itself in the ordinary field and woodland struggle for existence.

This observation was the key to understanding not only *Tradescantia virginiana,* but many other things which had been puzzling. *Tradescantia virginiana* in European gardens was no longer *Tradescantia virginiana.* As soon as a plant of true *virginiana* was grown in the same garden with a plant of *T. pilosa* there was a chance that pollen from a late-flowering *T. virginiana* might be interchanged with an exceptionally early-flowering *pilosa.* The seed pods of Tradescantia are small and inconspicuous and explode their seeds as soon as they are ripe. A hybrid seedling between these two species is somewhat more vigorous than either parent. Hybrids between the two species keep their leaves green all the year like *pilosa* but have flowers which are nearly as large as those of *virginiana.* As soon as one of these hybrids had reached flowering size it would automatically have been encouraged for its better garden effect. It would have crossed back to the parental species and to other hybrids or with other species of Tradescantia which were brought into the collection. The offspring of hybrids are a variable lot of mongrels; from them a good gardener can select and reselect.

For much of this story we have historical proof. We know that John and William Bartram, the early Quaker botanists of Philadelphia, had brought *Tradescantia pilosa* in from the west and that it had naturalized itself in and around their garden in Philadelphia. We know that they and others sent this species over to English gardens. We know from the very specimens referred to by Linnaeus that both species were doing well in Europe. What would be more likely than that the crossing which went on so readily in my Saint Louis garden might take place in England? Nor would it have stopped there. *Tradescantia ohiensis* had also been sent to Europe. Given two hundred years and more, what might not have been bred out of these mixtures in a succession of gardens where anything softer in color, or longer in bloom, or less weedy in aspect would be most likely to be saved? That is what *Tradescantia virginiana* of European gardens really is. It is a set of mongrels bred out of these three species. To judge from appearance and behavior, it has perhaps as much *pilosa* in it as *virginiana,* and more *virginiana* than *ohiensis,* but all three are certainly there and the mongrel types resemble known hybrids I have produced experimentally in my own garden. The change did not take place all at once. Mongrel blood entered so slowly into the *Tradescantia virginiana* of Europe that science had no record of what was happening. Plants which were only one-half, or three-fourths, or seven-eighths *Tradescantia virginiana* still bore that label in botanical gardens and nurseries. These mongrels were better garden plants and gradually they replaced the true species, but so slowly and inconspicuously that if any gardener or botanist ever noticed that the true *T. virginiana* had disappeared completely, there is no record of his discovery.

The irregularities reported in garden Tradescantias by the microscopists and which I failed to find in wild *virginianas* are of exactly the sort which are known to be frequent among hybrids. The kind of evolution which is going on in a garden is technically different from the kind which takes place in an ordinary woodland. What must be the complexities of our common weeds and cultivated plants? After three hundred years in English gardens it was only by persistence and good luck that the Tradescantia story could be worked out. Had they been in cultivation for thousands of years it would have been a much more difficult problem to interpret.

For Tradescantia the world's botanists respect our monograph. Wild-growing plants of *Tradescantia virginiana* are now correctly so designated. But in the world's gardens, a plant which traces back to that species only in part is still referred to as *T. virginiana* in horticultural catalogues and books about gardening. In virtually every botanical garden in Europe, if the plant is grown at all, it is still labeled *Tradescantia virginiana.* Any botanist or gardener who has known Tradescantia only in botanical gardens and perennial borders, will inevitably think, if he comes upon the true *Tradescantia virginiana* growing wild, that here is a new species which has little or nothing to do with *Tradescantia virginiana.* One naïve but kindly botanist who knew nothing about Linnaeus's authentic specimens even wrote Dr. Woodson and me a friendly letter to tell us that what we were calling *T. virginiana* could not possibly be that species since he had known *virginiana* since boyhood as a garden plant and it was something quite different!

Imagine what a more tangled scheme nature and man have woven with our older cultivated plants! Small wonder that it has been so difficult to name them precisely, that many botanists have given up the attempt and confined themselves to the simpler problems presented by wildlings from woodlands and mountains and seashores. In answering the question "What is *Tradescantia virginiana?*" we have learned how critical we must be of scientific names of plants even when they are used by scientists, and what kind of complexities we may expect in considering the relationships between man and plants.

III
The Greater Paradox

THE GREAT PARADOX that our commonest plants are the least known has given rise to a greater one, that this perilous situation is very generally unsuspected. Officially it is still within the province of taxonomy, the science of classification, to deal in a precise and scholarly way with the naming and classification of all plants, wild or cultivated. The earliest taxonomists did indeed concentrate their attention on everyday plants. So gradually did their successors become aware of the special difficulties posed by the plants of fields, gardens, and dooryards that they drifted slowly into their present policy of avoidance. So far as I can find out no complete public review of this position has ever been made by themselves or by others. Between taxonomists and other scientists there is now enough of a gulf so that most of the latter assume the classification of cultivated plants to be at least tolerably well understood. Nothing is farther from the truth. The examples of the bearded irises and the cultivated spiderworts described in the previous chapter are closer to being the rule than the exception. Most modern taxonomists do next to nothing with cultivated plants; many deliberately avoid studying or even collecting them. As a result the scientific botanical name affixed to most cultivated plants becomes just an elaborate way of saying "I do not know." It is a big inaccurate pigeonhole under which many kinds of facts are thrown together helter-skelter. Because the name is Latin and looks scientific, it is treated with the respect our technical civilization reserves for matters which it does not understand.

This is a serious business. To the rest of the scientific world, to agriculturists, to anthropologists, to plant breeders, to geographers, to economists, to historians, the accurate classification of the plants most closely associated with man is of more importance than that of all the other plants of the world put together. To the average taxonomist, these are the very plants least effectively dealt with by his methods; he prefers not even to think about them.

There are logical reasons for this avoidance, as we shall see. As a matter of fact, most of this chapter is a defense of taxonomists, an attempt to understand why they so gradually found methods inadequate for the plants of man's transported landscapes. First of all, though, it may be profitable to cite a few specific examples of the attitudes of modern taxonomists towards the classification of cultivated plants. My own office is surrounded on three sides by an herbarium presided over by my colleague, Dr. R. E. Woodson, one of the outstanding taxonomists of the United States. Ours is a privately endowed institution; expenses are going up, income is coming down. The steel cases of the herbarium already fill approximately every corner of the building. There are no funds for expansion; there is little spare room left in most of the cases. Dr. Woodson and his staff are doing the intelligent thing. They are going through the herbarium, case by case, and taking out everything which may reasonably be spared. All the duplicate specimens are coming out, to be exchanged or sold to other institutions. Specimens lacking complete data to make them of scholarly value but otherwise in good condition are being laid aside for teaching purposes here and elsewhere. In addition, the majority of the specimens of cultivated plants are being discarded. No other herbarium wants most of them, and they are being thrown away; all the kinds of cultivated chrysanthemums grown at our own garden half a century ago, a large collection of old-fashioned varieties of cannas, a collection of varieties of the cultivated grape made in the middle of the nineteenth century by Dr. George Engelmann, the first curator of the herbarium. No criticism of Dr. Woodson's policy is intended or implied. He is a modern taxonomist, making the best use of the limited available assets. It does demonstrate, however, the extent to which the classification of cultivated plants is deliberately avoided by modern taxonomists.

Sometimes it is hard for taxonomists even to realize that other biologists can be really concerned about the identification of cultivated plants as another example will show. Some ten years ago I became interested in the classification of maize, the crop which we Americans began by calling Indian corn and now refer to simply as corn. Virtually nothing was then known about its classification, how many major kinds of maize there were, and how they might be distinguished. I set to, enthusiastically, getting together as much information as possible about these matters. One of my taxonomic friends, a most obliging person, was spending some months in the mountains of Latin America with a student assistant. I told him how much I would like specimens of maize and he faithfully promised to bring me some. I emphasized how much I would like some from the mountains and assured him that if he would just press one or two corn tassels, as if they were ordinary grasses, that I would be able to use them effectively. On his return I inquired if he had found any maize. "Why, Andy, you are mistaken. There isn't any maize growing wild anywhere there. All we saw was just growing in *fields*." He had so little interest in cultivated plants that he could not even imagine that anyone else could use an herbarium specimen of a plant growing in a field! To comfort me in my disappointment he said, "Oh well, it wasn't anything you would have wanted; it was just like our ordinary corn." But the student assistant interrupted him.

"Oh no, Doc, don't you remember? When we camped beside that cornfield on the volcano I called your attention to the fact that it was funny-looking stuff with coppery red leaves." Now the "funny-looking stuff" with the coppery red leaves was precisely one of the problems connected with the history and classification of maize which I was trying to run down. There are bright-colored kinds of maize at high altitudes in Mexico and greener ones in the lowlands. There are bright-colored varieties in highland Peru and greener ones along the coast. I would have given a good deal to have had just one specimen of that funny copper-colored stuff. Yet if his assistant had not been young and uncorrupted I might never have known even that it was there! My friend has helped me in various ways: he keeps promising to get me specimens from his old friends in that part of the world, but ten years later I am still trying to get specimens of the copper-leaved maize from that volcano.

The final example is even more significant because the scientist whose attitude is analyzed, Dr. E. D. Merrill, is certainly our most eminent American taxonomist, and a scholar of great ability and world-wide reputation. He furthermore, unlike most of his fellow taxonomists, has taken some interest in cultivated plants, and has written several technical papers on the origin and distribution of economically useful plants in southwestern Asia. His knowledge of the flora of Asia and the South Pacific is encyclopedic; he has the reputation of being able to name more species at sight than any other American taxonomist. Consequently there come to him many unnamed collections of pressed specimens from the Oriental tropics. In the days when he was a busy administrator and had to fit this activity into a full day he did the naming with the aid of one or more assistants. He would pick up a specimen, reel off the complete name, or if he did not know that, the name of the approximate group in which it belonged, and hand it over to an assistant and go on to the next specimen. If the material was being incorporated into the herbarium of which he was in charge, whenever he came to one of those plants which have associated with man for so long that no one at present knows from whence they came, he would mutter "Pantropic weed," and throw the specimen into the wastebasket. Some other curators, lacking his forthright disposition, kept such specimens in their collections, but their attitudes were virtually the same. One of my own students, a brilliant young taxonomist, when pressed for a definition of a weed, quickly replied, "An ubiquitous plant, with variation you can't get your teeth into, which clutters up herbaria."

One could cite scores of such reflections of the general attitude of first-rate modern taxonomists towards cultivated plants and weeds. Because these attitudes have never appeared in print, only an occasional biologist who is closely associated with taxonomists realizes how far this trend has gone. It is not yet common knowledge, even among scientists, how completely the classification of cultivated plants and weeds is being avoided. Therefore, before this story of plants and man can be continued, we shall have to take a long detour and explain the perfectly natural way in which the stagnation came about. I shall have to describe how taxonomy got started, how it gradually evolved its techniques, and why it is that in the middle of the twentieth century we wake up to find many of our dooryard plants virtually unknown.

If you want to know something of these beginnings, go to the treasure room of any big library and study one of the herbals of Leonhard Fuchs of Basel. Fuchs was one of those Renaissance pioneers who forsook the library gloom of the Dark Ages to study plants themselves instead of copying copies of copies of ancient manuscripts about plants. A splendid woodcut at the end of the volume pictures him as a big, broad-shouldered Henry-the-Eighth sort of man with handsome clothes and a general air of getting things done. The folio edition of Fuchs is pleasant to leaf through. It is a big volume about the size of the thick dictionaries which used to be standard equipment in elementary schools. On nearly every left-hand page there is a woodcut of the plant under discussion and facing it, an account, in German black letter, of how it may be distinguished from other similar plants and a long catalogue of the ways it may be used.

The volume was produced under Fuchs's careful direction. We know that he supervised the illustrations; in addition to his own picture, there are woodcuts showing his artist at work drawing the original specimens and his engraver industriously transferring these drawings to wood blocks. In Fuchs's big herbal, as in all the botanical work of his day, there is no distinction between garden plants and wildlings, between the ubiquitous weeds of the dooryard and the flowers of heath and woodland. The common cabbage is drawn with as loving care as any illustration in the book. I doubt if any picture of a cabbage has ever charmed more people; it is worth going and drawing the book out of the library just to dwell upon the crisp arching lines which make so very cabbagey a cabbage stand out on the big white sheet.

To go on and show how taxonomy developed from these modest beginnings let us walk southwards down the High Street in Oxford and, just short of the bridge, turn into the old botanical garden opposite Magdalen College. There one can find remnants of a typical European botanical garden, a garden primarily for study and not for show. There are narrow beds with grass walks in between them and the whole garden is protected by high stone walls. It was in a botanical garden of some such sort that Leonhard Fuchs and other herbalists got to know their plants so intimately. This Oxford Botanical Garden in the river meadows across from Magdalen has sheltered a succession of men who loved and studied plants. The greatest of those who worked there, as well as one of

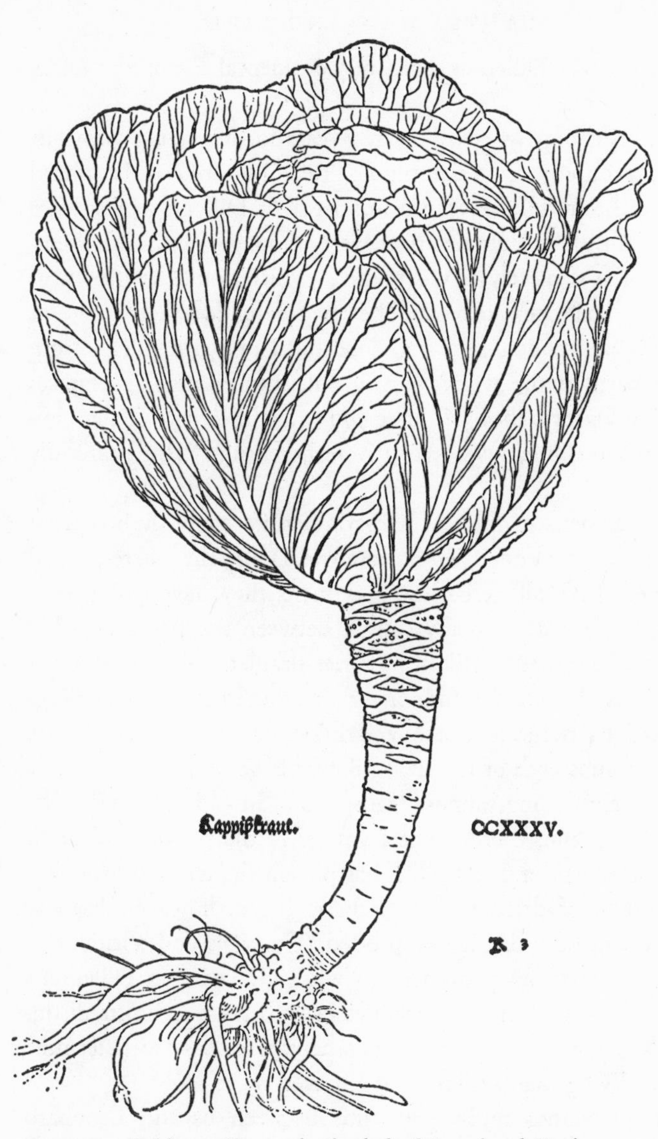

FIGURE 3. Cabbage. From the herbal of Leonhard Fuchs, 1543

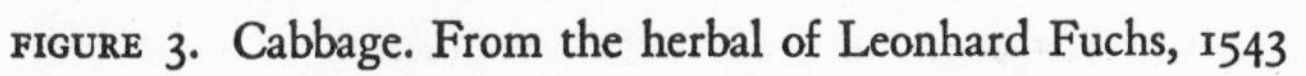

the earliest, was Dillenius, born in Continental Europe as Jakob Dillen.

Dillenius brings us two hundred years farther along with our story of taxonomy, from the middle fifteen hundreds of Leonhard Fuchs to the middle seventeen hundreds of Dillenius's residence at Oxford. By that time botanists were no longer confining themselves to the relatively few specimens which could be grown in their own little garden plots. They had extended their range of vision by assembling herbaria, collections of dried plants from various parts of the world. If you will inquire at the botanical garden in Oxford, they will take you to another part of the University where Dillenius's own private herbarium is still carefully preserved.

The specimens in this old herbarium differ only slightly from those in a modern one; two hundred years ago they were pressed and dried. Like all herbarium specimens they have the general look of a plant accidentally caught between the pages of a big book and kept there until all its juices drained away. The flower of an iris is all collapsed into one plane, the flutings and rufflings of its corolla reduced to a few wrinkles and marginal scallops. The crisp substance of the living flower is gone. It has become a fragile papery thing, almost transparent. In old specimens like those of Dillenius there is not a vestige of the original color; all the plant, flower and leaf alike, has put on the characteristic dull dead brown of dried leaves. Its basic shape, however, has not changed since the day it was pressed. The sizes and positions of the tiny hairs on its sepals are just what they were in Dillenius's day. Most of the little technicalities by which taxonomists distinguish between one species and another are as easy to determine as in the living plants, some of them even more so.

Dillenius stands midway in time between us and Leonhard Fuchs. In this last two hundred years taxonomists have increasingly had less and less to do with gardens and more and more to do with dried specimens. One could almost say that modern plant taxonomy has become the science which deals with herbarium specimens. Since the time of Dillenius, their use has expanded more than it has developed. That is to say, taxonomists work with many more herbarium specimens than in his day but their techniques are virtually the same.

A good general herbarium has hundreds of thousands, if not millions, of such specimens. Nowadays they are mostly filed in big steel cases about the size and shape of an old-fashioned German wardrobe. In a modern herbarium these cases stand in row after row, with here and there a group of table-height cases to provide a working surface. The actual specimens are filed closely together in big Manila folders on shelves in the steel cabinets. Each specimen is mounted on a sheet of heavy white paper of standard size about a foot wide and a foot and a half long. In one corner, usually the lower right-hand one, is a small label telling where the specimen was collected and when and by whom. Frequently a special number is added to make precise reference to this collection a simpler matter in scholarly publications.

Though they have sometimes been contemptuously referred to as "taxonomic hay" by other biologists, herbarium specimens can be quite romantic in their own dry way. Let us take a Manila folder out of the case at random. It happens to hold specimens of a little grass from southern New Mexico. Here are two herbarium sheets collected in the mid-nineteenth century by the naturalist who rode along on the survey of the Mexican boundary carried on under government direction. Here are sheets collected at about the same time by Wislizenus, one of those fiery German intellectuals who swarmed to the New World after the unsuccessful German liberal movements of the early nineteenth century. Here is a sheet collected by a woman whose husband was a mining engineer stationed on a remote mountain range. She turned their isolation to account by making a study of the native flora. Here are several elegant sheets prepared by a Boston Brahmin, who for thirty years tempered life in Boston with discreet private explorations in the Southwest. Here is a new sheet from a scientist at Los Alamos who has turned to natural history for recreation. Here are others collected twenty years ago by gentlemen who are now sedate heads of botany departments in various American universities, but were then eager young students who scraped enough money together to buy old secondhand automobiles and tour. the Southwest for a summer, selling enough specimens to botanical institutions to defray most of their expenses.

The herbarium method is one of those simple techniques which are more important than they seem to be from casual inspection. In the first place, it works directly from documents instead of from interpretations of documents. In the language of the historian, it is a technique based solely on original sources. When a modern taxonomist monographs a group of plants, a considerable part of his published work may be the detailed citation of the specimens he examined. His belief that certain of these specimens all belong in one species is an idea, possibly right and possibly wrong. The specimens themselves, however, are a set of facts — good hard, solid facts. They are more than a reflection of reality; they are reality itself. If one wants to get behind the monographer's idea and look at his facts, one has only to go to a good herbarium, look through the proper folders, draw out the duplicates of some of the collections he cited, and under one's own eyes are the very facts on which the monographer based his judgment.

Like any technique based on objective data, the herbarium technique is therefore scientifically sound. It has another feature, also illustrated in the example just cited: it is standardized, efficient, and international. Taxonomists, though frequently considered hopelessly conservative, were really among the first scientists to seize upon the efficient principle of standardized interchangeable parts, or rather, they invented it for themselves before its advantages had become apparent to the industrial world. Throughout the world, herbarium equipment is virtually the same. Taxonomic institutions in Asia, Africa, Europe and America, make the same kinds of collections. They can be stored together, filed together, sold, borrowed or traded from one institution to another. Collections made by the Indian Forest Service, by the Michigan Biological Survey, by the Arnold Arboretum Expeditions to western China, by the summer-school students of the University of Wyoming, all go into the same international hopper. Collections can be made in duplicate or triplicate or in sets of ten, if the plant is common enough, so that at London and Paris and Cambridge and Saint Louis there will be interchangeable vouchers of what kind of plants were growing in that particular place.

Like any technique, however, the herbarium method has its limitations, and its dangers. Techniques tend to become ends in themselves instead of means to an end. Building up an herbarium appeals to that simian strain which is in all of us, that instinct to hoard for the fun of hoarding, which is the basic impulse behind attics, stamp collecting, the treasure rooms of libraries and private fortunes grown beyond the capacity of increasing usefulness to the owner. Herbarium administrators tend to become stamp collectors instead of biologists. They frequently become more interested in their herbarium specimens than in the biological usefulness of the collection. Dr. Merrill and a few other progressively-minded taxonomists have demonstrated how the scientific value of an herbarium can be greatly extended by augmenting the actual specimens with detailed notes and by including various kinds of printed and typewritten information side by side with the herbarium sheets. Such suggestions distress the stamp collectors. Like old-maid librarians they are primarily interested in having neat collections on their shelves, and only secondarily in the uses to which the collections are put. Dr. Merrill's innovations, in spite of his demonstrated eminence as a taxonomist, in spite of his vigorous and challenging personality, have been adopted by few curators. Some have gone further and have even discarded the detailed notes on specimens he sent them because the inclusion of such extraneous material destroyed the chaste uniformity of white sheet and printed label.

To this tendency, present in all herbarium-minded taxonomists, is added the overwhelming fact of the herbarium specimens themselves. It takes a lot of specimens to make a good herbarium and it takes a lot of time to care for a lot of specimens. Whatever the direction of his natural tastes, a modern plant taxonomist is going to have to spend a good portion of his mature life with herbarium specimens. Not only must they be collected and mounted. They must be exchanged, lent, borrowed, and fumigated as well as studied. No first-class herbarium has less than a half million and our largest herbaria run into several million sheets. On the whole the ablest taxonomists work in the largest herbaria; by the duties of their positions, their time is taken up with the acquisition and protection of herbarium specimens. As their professional lives develop they tend to think and write and speak more and more exclusively in terms of herbarium specimens and herbarium techniques. At meetings of taxonomists it is these leaders who set the tone. Younger men not already herbarium-minded through professional necessity, voluntarily become so through professional pride.

All of these facts have a bearing on the question of why we know so little about cultivated plants and weeds, but the most important fact has not yet been brought out. It is simply that the herbarium method works so wonderfully well with everything *except* those plants associated with man. With the plants of mountain and jungle the herbarium worker has done astonishingly and increasingly well. Species are but judgments as we have already indicated but so excellent is the herbarium method that there is usually very general agreement between taxonomists as to how the genuinely wild members of a flora ought to be classified. Two monographers in different parts of the world, working with the same group of species, may independently come to surprisingly similar conclusions as to how many species there are, how they may be distinguished and in what genera they should be classified.

An example will show how an herbarium monographer operates. Since we have already discussed spiderworts it may be simpler to take an illustration from that genus. When Dr. Woodson and I were monographing these American Tradescantias, a minor problem was presented by a variable slender-leaved lot of plants, varying in size, number of flowers, number of branches and many other features, and commonly found in sunny places all the way from the Rocky Mountains to the eastern seaboard. There were obviously various strains and substrains. A considerable number of names had been suggested for them at one time or another; there was apparently more than one species but how to separate them? Characteristically the plants were glabrous but many of them had a few hairs on or near the flowers. It required only a casual inspection with a hand lens to find that in some specimens the hairs tapered to a fine point, that in others the hairs were shorter with a glandular knob at the end. Sorted out on this basis the glandular-haired ones were found to come from the Great Plains and the edge of the Rocky Mountains, those without glands from the Middle West and the Southeast. With this to go by, the apparently hairless specimens were re-sorted geographically, those from the Great Plains in one pile, those from the East in another. It was then found that very few of these specimens were absolutely hairless. Sometimes whole flowers or even flower heads would be without a perceptible hair but others could be found elsewhere on the plant. But in these geographically sorted specimens it was easy for us to see that in the Western group if there were but one or two hairs they would be at the base of the flower, while in the Eastern ones, if there were only a few, they would be at the very tip of the sepal. By working back and forth in this way from differences in the kinds of hairs to differences in distribution of the plants and then back again to differences in hair position, it was possible in a very short time to discover that we were working with two widespread species. There were other characters, less easily put in a few words, by which the two sets of specimens could be distinguished. The Western ones had smaller flowers, fewer joints on the main stem, more slender bracts beneath the flowers, and harder, darker leaves. The two species occupied different areas and though each was greatly variable they had a core of common differences and could readily be described in such a way that other scientists could classify future collections using our keys and plates.

Our specimens had been largely from habitats not greatly disturbed by man. Though a good many were from country roadsides, very few were actually from gardens. Had someone presented us with an equally large collection made exclusively from gardens, vacant lots in cities, dump heaps and deserted fields, we could not have worked effectively. There would have been no geographical continuity of related plants. Cultivated plants spread here, there and yon as their seeds or bulbs or rhizomes are given and traded and sold. I have already described in an earlier chapter what three hundred years of cultivation have done to *Tradescantia virginiana*. Freak forms with albino flowers or seedless fruits or feathered leaves are seized upon and widely propagated whereas in nature they would have been quickly eliminated. Species which seldom or never had the opportunity to hybridize are brought together under conditions where many seedlings are raised and the off-type ones have, if anything, the better chance to survive.

As we look back now, it is easy enough to see why cultivated plants are of another order of taxonomic difficulty than are the genuinely wild members of a flora. Taxonomists, shifting slowly from reliance upon their medieval botanical gardens to their modern herbaria, took generations to drift from a special interest in the plants used by man to an almost complete disregard for them. There is a century between the time when Dr. Engelmann, an outstanding taxonomist of his day, lovingly made a set of specimens of the grapes cultivated in Missouri and the time when his successor, an outstanding taxonomist of our own times, despairingly threw them away.

To this gradual neglect of cultivated plants the collector and field taxonomist has contributed quite as much as the herbarium specialist. Once he is in the field, the average taxonomist is an incurable romantic. Watch him take a group of students on a field trip. The nearest fragments of the original flora may be miles away and difficult of access but that is no barrier. With truck, bus, train, jeep, or car, on foot if need be, the class is rushed past the domesticated and semidomesticated floras among which they spend their lives to the cliffside or peat bog or woodland which most nearly reflects nature in prehuman times. Or follow in the taxonomist's footsteps when he leads an expedition into the tropics or establishes a field station in Central America. There one meets with wide areas of thorn scrub and savanna used as range and pasture, considerable land in field crops and in gardens, and at higher elevations, wide expanses of pinewood, more or less pastured and more or less cutover. Very rarely a remote ridge rises to a peak and is clothed around the summit with a cloud forest.

From the moment a taxonomist arrives in the area these tiny patches of cloud forest are the center of his interest; from his behavior one would suppose they were his main reason for being in Central America. Admittedly they are beautiful and biologically interesting in more ways than one. Scientifically they offer no more fundamental problems than do dump heaps or dooryards or maize fields or village gardens, all of which will be ignored by your true taxonomist.

This appetite for virgin vegetation rises directly from some deep emotional urge rather than from cool professional competence. When the taxonomist mentions the cloud forest, when he makes his plans to go there, when he climbs up the lower slopes of the peak, his eye lights up and his voice has a richer vibrato; obviously he is somehow in love with the cloud forest. I do not mean to suggest that it is not awe-inspiring to any normal human being. Far from it. The shifting cloud mists and the heavy vegetation produce an atmosphere which is virtually submarine, with a ghostly green light. The towering trees, the lianas, the translucent filmy ferns and mosses, dripping with moisture, shut one off from the sky above and the earth below. It is an experience to remember. The religiously-minded person moves quietly or stands open-eyed in the presence of something greater than himself with which he is somehow in communion. But not your taxonomist. He rushes about with a great excess of energy, throwing the plants into presses, searching here and there for something yet uncollected. It would seem as if somehow he is trying to make the cloud forest a part of his professional self, to get the forest into his grasp. He is in such an itch to return again and again or to find another cloud forest that the special problems of those plants associated with man are kept outside his area of thought.

But are there no exceptions? Only enough to prove the rule. Take any ordinary taxonomist and make him responsible for an herbarium of cultivated plants and he only gives his job lip service. He is not happy about it. He realizes how inadequately he is dealing with such problems and escapes into the study of native floras whenever he can. The only professional American taxonomist to concentrate upon cultivated plants and to urge that their taxonomy be studied has been L. H. Bailey. Almost singlehanded he built up the Bailey Hortorium, an herbarium devoted to the classification of cultivated plants. It is only he and a few others who were forced into such work by the nature of their jobs who have kept the naming of cultivated plants from being complete chaos.

As we shall see in another chapter it is now getting to be perfectly possible to deal effectively with cultivated plants by simple extensions of the herbarium method. One has to take a special interest in the problem, to be intelligent about the matter, to supplement the customary specimens with notes and photographs, to concentrate upon particular crops or particular regions. Two serious deficiencies of the herbarium method as it is now practiced could be remedied by industry and ingenuity if the urgency of such improvements were more generally recognized. (1) Nearly any ordinary specimen is an imperfect and incomplete reflection of the living plant from which it was made. If the plant is over three feet high it is obvious that only certain selected portions can be affixed to a sheet a foot and a half long and a foot wide. Making a good herbarium record of tall prairie grasses is something like trying to stable a camel in a dog kennel. This difficulty can be overcome by combining specimens of certain critical portions of the plant with photographs, diagrams and notes, but that is seldom done. (2) An even greater deficiency of the ordinary specimen is that it is just one individual. It tells us nothing about the plants which are growing with it. An herbarium specimen of one plant can be accompanied by statistics or even by specimens of some critical portion of the plants growing with it. By such minor extensions of herbarium techniques it is possible to work effectively with cultivated plants and weeds. They tend to be ignored in modern herbaria, not because that is really an ineffective place to study such problems but because taxonomists drifted gradually away from a real interest in what had originally been their main job.

Such a neglect of a socially and economically important field is not unusual in science. Science for all its integration of fact and theory is a strange kind of anarchy. There is little over-all planning. Discoveries are made not because there is a crying need for knowledge in that area but because someone has a fascinating new technique and young men become intoxicated with the new field of exploration which has been opened up and dash off into it. There are fads in science. A problem which looks humdrum gets passed up for one in which more scientists are currently interested. Eventually someone goes back to the neglected subject with a new set of ideas. That is just now beginning to happen with the study of cultivated plants.

I V

The Clue from the Root Tips

SCIENCE, as I have just been saying, is an anarchy in its general lack of over-all planning. If taxonomists (classifiers by definition) gradually shift away from any serious concern for the special problems of cultivated plants, they are none the less reputable scientists and still worthy of public support. But so anarchical is science that, if taxonomists do so neglect a field which was once their own, there is no way of stopping any other groups from moving in. The last few decades have seen such a movement, one that began so slowly and on so many fronts that it has not until now been defined as a trend. Though my own trespassing in this field may have been conspicuous it was far from being the first or the most important. Trespassers we all have been in a very real sense; none of us was well grounded in taxonomic techniques, some of us began bunglingly with the assumption that there were no techniques to learn or at least none worth learning, all of us came in from the outside. Plant breeders, geneticists, cytogeneticists, ethnobotanists, geographers, for one reason or another we found a need for understanding the classification of cultivated plants. Gradually and independently there came a realization that professional taxonomists scarcely concerned themselves with such matters any more.

Of all those who rushed in from the outside none arrived with stranger new evidence than the cytologists. Cytologists are microscopists; they study plant and animal cells under high magnification. Their commonest technique has been to take some easily accessible portion of a plant or animal (in plants it is frequently the tip of a rapidly growing root) and immerse it in various successive solutions to preserve and stain its essential features and then to study these pickled remains enlarged a thousand times or more under the microscope. Who would have supposed that men of this sort would have new clues to the origin and history of the plants so closely associated with man? Yet it is they who, though generally ignorant of prehistory, knowing little or nothing about taxonomy and usually so scornful of its innate conservatism that they did not wish to learn any more, have made the most startling discoveries in this field in the last few decades. It is they who have produced exact evidence as to which kinds of primitive grasses were combined in the Stone Age to produce our wheats. It is they who can prove without the shadow of a doubt that the Asiatic cottons (perhaps the wild ones, perhaps the cultivated ones) somehow crossed the Pacific or traveled around it by slow stages and played a definite role in establishing American cultivated and weed cottons. It is they who narrowed down the problem of where tobacco might have originated and explained such modern miracles as the loganberry. Nor have they been content merely to launch fantastic hypotheses; by further developments of their techniques and with the help of plant breeders they have been able to re-enact these hypothetical histories and actually (as we shall see) to recreate such crop plants *de novo* from their primitive ancestors. Brilliant as this new evidence is, it tells us about only one or two details in the origin and development of certain crops and weeds — nothing about the rest of their histories, and nothing at all for many other crops and weeds. If we revert to our previous conception of the history of these important plants as a complex detective story with many kinds of clues, then cytology furnishes a disconnected set of brilliant flashlight photographs, illustrating with tantalizing clarity just one or two phases of the mystery.

To understand the nature of the cytological evidence and why it may have such a fundamental bearing on these important problems it will be simplest to return again to our spiderworts and join a cytologist in studying them under the microscope. If we take a spiderwort which is in flower and choose a bud whose petals are just beginning to gain their characteristic blue purple, we can open it with a needle and drag out the rapidly maturing stamens. It is the stamens which, had the flower opened, would have burst along their edges exposing that golden dust, the pollen grains. In these immature buds the pollen is now rapidly maturing. The nuclei of the grains are in the process of dividing to form germ cells. So large is the spiderwort nucleus, so clear the sap by which it is surrounded, and so readily does it take up stains that here is the best material in the world for studying the complicated system by which cells divide and create new cells. Midway in this process the nucleus as such has disappeared and the deep-staining hereditary material has shortened and condensed into a small number of definite bodies called the chromosomes. In the simplest spiderworts there are just six of these, six short dark ribbons, each one slightly constricted in the middle so that it looks like a rather elongated bow tie. Were we to examine the cells of the root tip we would find not six chromosomes in each cell but exactly twelve, and if we were to make an exhaustive study of their slight differences we could prove that the twelve are made up of two sets of six. If we study other plants of the same strain of spiderwort, we usually find the same result, six chromosomes in the germ cells, two sets of six in the other tissues. This becomes understandable when we study the whole life history of a spiderwort; an egg cell with six chromosomes is fertilized by the male germ cell which also has six chromosomes, and the resulting plant characteristically has six plus six or twelve chromosomes in each cell until it halves that number in the complicated process of producing germ cells.

Nor is this process unique among spiderworts. There is a definite number of chromosomes in the cells of lettuce, of oak trees, of begonias, of elephants, of angleworms, of June bugs, of mice, and of man himself. Under normal conditions, in all these kinds of plants and animals there are two sets of chromosomes in most of the tissues, and one set in the germ cells, but the numbers and sizes and shapes of chromosomes vary greatly from one kind of plant or animal to another.

To explain what this kind of evidence has to do with determining the relationships of Asiatic and American cottons, we can do no better than to describe in considerable detail a modest little primrose, *Primula kewensis.* It was in demonstrating the nature of this hybrid, that some of the first clear evidence for the origin of wheat and cotton and tobacco was brought to light. Most appropriately *Primula kewensis* originated in England. It is in England more than any other part of the world that the common yellow primrose grows in greatest profusion along the roadsides and about the hedgerows. It illuminates the misty English springtime with the glory of its massed flowers of a yellow so perfect in its clarity, so affecting in its simplicity, that one has no other way of describing it than to say that it is a true primrose yellow. The English, having associated since childhood with these lovely blooms, have extended their affection to all primroses and a goodly number of the Primulas of the world have been brought to their gardens and greenhouses. Two of these exotic primroses, one from the Himalayas and one from southern Arabia, were growing at the Royal Botanic Gardens at Kew where they were hybridized by an alert gardener, producing a vigorous hybrid which before long was named *Primula kewensis* after its birthplace. It was easily grown in the greenhouse and it flowered well in the short days of the English winter, bringing a promise of spring and wild yellow primroses to housebound Englishmen. For these reasons it had distinct commercial possibilities and a special effort was made to propagate it. Unfortunately, like many hybrids between distinct species, it was quite sterile; it could only be increased by the slow process of growing the original seedlings until they became big plants, dividing each of these into several pieces with a sharp knife, potting them up and then growing these little ones up into big ones and dividing them again in due time. However, on at least two occasions one of these sterile hybrids threw up a lush, darker green branch with wider, thicker leaves and slightly larger flowers, which not only set seed, but set it in abundance. Furthermore when this seed was planted, the seedlings, unlike those of most hybrids, were not a swarm of varying mongrels. They were similar to one another, they resembled the giant branch from which they came, they were themselves fertile, and they gave rise to a fertile and commercially profitable race of *Primula kewensis.*

It was the cytologists who eventually solved this mystery by counting the chromosome numbers in the sterile hybrid, in the fertile branch and in the original species. The sterile hybrid had nine chromosomes in its pollen grains and nine pairs of chromosomes in its root tips. So did its Himalayan parent, *Primula floribunda,* and its Arabian parent, *Primula verticillata.* The fertile branch, however, had eighteen pairs of chromosomes in its tissues and eighteen chromosomes in its pollen grains. If we think of the chromosomes in these primroses as coming in sets of nine, then the parental species and the sterile hybrid had two sets in most of their tissues, and one set in each germ cell, while the fertile branch had four sets in most of its tissues and two sets in its germ cells.

Now that many other plants have been found in which such changes occurred (and particularly since plant breeders have learned how to produce them artificially) it is convenient to have special names for this phenomenon. The sterile hybrid and its parents are said to be diploid, that is, having two sets; and the fertile branch is tetraploid, that is, having four sets. We may conveniently, for purposes of discussion, merely lump it with all those organisms having more than two sets (three, four, six, eight, ten as the case might be) and call it a polyploid.

The fertile branch on the sterile primrose hybrid was a polyploid branch. It probably originated when the nucleus of a cell divided, thus doubling the chromosome number, and for some reason the cell itself failed to divide. This single polyploid cell divided in a perfectly regular fashion and produced two polyploid cells and eventually gave rise to a whole mass of polyploid tissue and at length the polyploid branch. In these primrose hybrids, polyploidy turned a sterile diploid into a fertile tetraploid. In the sterile diploid there had been two sets of chromosomes, one from *floribunda* and one from *verticillata.* When they came to flower and produce germ cells the two sets were too unlike to go through this complicated process successfully; the resulting germ cells aborted before they were mature and the hybrid was sterile. In the fertile branch, however, there were two complete sets of *floribunda* chromosomes just as there are in *Primula floribunda* itself, and two full sets of *verticillata.* Each of these set pairs could go through the complicated process of halving its number virtually as well as in the original parental species. Every pollen grain therefore was supplied with one set of *floribunda* chromosomes and one set of *verticillata.* All the egg cells were similarly equipped. As a result, fertility was restored to practically that of the original species and the hybrid was virtually true-breeding since all the pollen grains and all the egg cells were similarly

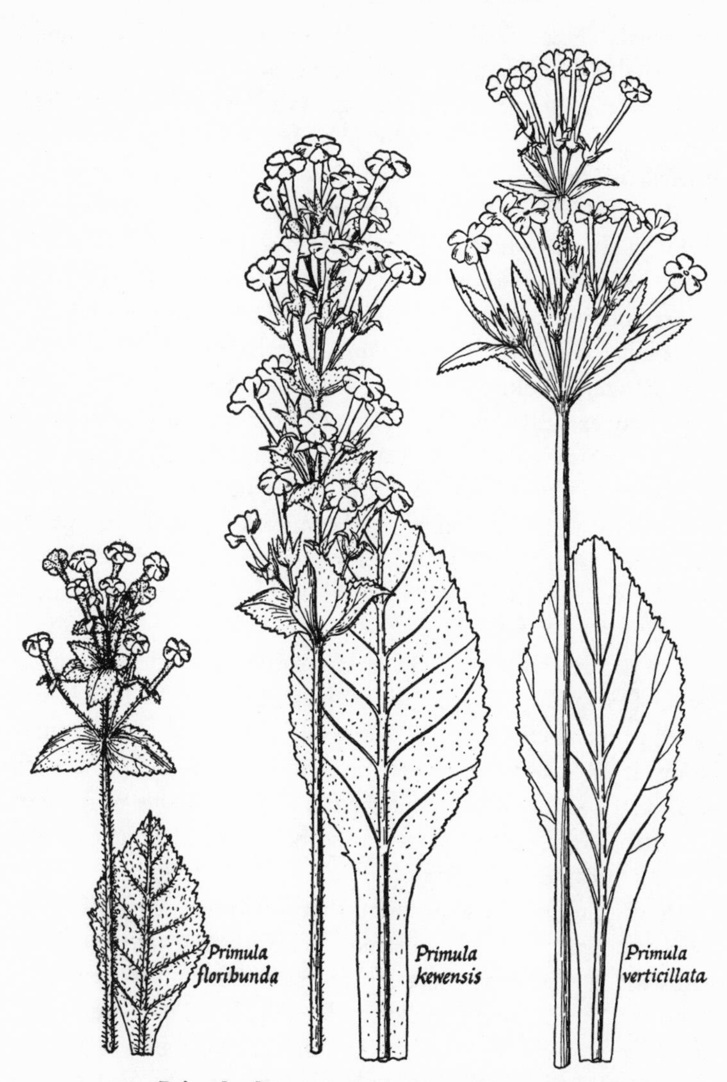

FIGURE 4. *Primula kewensis* with its parents, *Primula floribunda* (left) and *Primula verticillata* (right).

equipped. The original diploid hybrid *kewensis* was like a mule, it was half a *floribunda* and half a *verticillata* and these two halves were too unlike to function together in complete harmony. The polyploid hybrid was essentially a whole *floribunda* plus a whole *verticillata*. Polyploidy, therefore, provides a means by which a sterile hybrid may at one jump become completely fertile, and virtually true-breeding, and start off on a career of its own.

We do not yet know exactly what percentage of cultivated plants and weeds have originated as true-breeding polyploid hybrids but it must be a pretty high one. Polyploidy is not the only way in which cultivated plants arise, nor does it ever play an absolutely exclusive role, but it is one of the most important processes in the development of man's transported floras. Now that we understand polyploidy and can sometimes even produce it at will, many puzzling things in the history of cultivated plants are easier to understand. Take for instance the loganberry, one of the few important crops which have originated in modern times.

The loganberry turned up in the garden of Judge Logan of Santa Cruz, California, in the eighteen-eighties. At first it was taken to be exactly what the Judge said it was and what we now know it to have been, the result of a cross-pollination between the west coast blackberry and the red raspberry. Due to its great commercial success and to its almost miraculously sudden appearance, it became the object of a good deal of study and of even more dispute. It was soon found that unlike ordinary hybrids it bred practically true. Furthermore, repetitions of the suspected cross did not produce more loganberries. The authorities of the time pontificated that it could not be a hybrid. There were some who were still convinced that it must be one but they were shouted down for thirty years by the experts who had not yet heard of polyploidy or, having heard, did not wish to revise their thinking. With the discovery of polyploidy one had only to count the chromosomes in the root tips to prove that the loganberry was a polyploid and therefore might well be of hybrid origin and yet true-breeding. The loganberry and other similar polyploid compounds of raspberries, dewberries, and blackberries, have now been carefully analyzed and there is no reasonable doubt but that the whole lot of them are true-breeding polyploid hybrids. With cytological control over the material it is now possible for plant breeders to produce such polyploid hybrids deliberately and an increasing number of them are coming on the market every decade.

Most of our cultivated polyploids are of much more ancient origin. Take the cultivated wheats. It had been known for years that the wheats of the world were more than just a large number of varieties of wheat. They belonged in three or more great groups; crosses between some of the groups were hard to make and such hybrids were more or less sterile. The cytologists showed that there was a very real background for these groupings. Einkorn wheat, an ancient cereal of Neolithic times, now practically disappeared as a crop plant, was a diploid; all the other cultivated wheats were polyploid. Some of them were tetraploid, including those protein-rich varieties which make a sticky flour and are known as "the macaroni wheats." The most important wheats of all, the bread wheats, were hexaploid, that is, they had six full sets of chromosomes. If each of the capital letters A and B and D designates a set of seven chromosomes, then the einkorns were of formula AA, the emmer and the macaroni wheats were AABB, and the bread wheats were AABBDD. Presumably the einkorn wheats had evolved out of the wild wheats of the Near East, since some of these were also simple diploids of the constitution AA, but where did the BB's and the DD's come from? This fascinating puzzle is not yet solved down to proving the last detail, but we are already reasonably certain of the general outline of the story and even have exact experimental proof for some of it. It is clearly apparent that the cultivated wheats did not spring from the wild wheats alone; the BB's and the DD's must represent other distinct genera of plants. BB may well be a quack grass, *Agropyron triticeum,* which is wild in the eastern Mediterranean region. A polyploid of it and einkorn back in Neolithic times may well have produced the first of the tetraploid AABB wheats from which emmer, the Persian wheats, and other modern tetraploids were eventually bred. The hexaploid (AABBDD) bread wheats were most probably produced from accidental hybridization between the AABB tetraploids, and a bristle-headed little weed of the Near East, *Aegilops squarrosa,* which supplied the necessary DD for the finished hexaploid.

By this time, the ordinary reader, showered with hexaploids and tetraploids, may begin to feel that the introduction of such Latin names as *Aegilops* for simple weed grasses is more than he should have to put up with. "Away with your *Aegilops,*" he says, "let us have the plain common name." Well, I might, if he so insists, use a common name for *Aegilops,* but before I replace *Aegilops* with something closer to ordinary life, let me remind you that these are not English or American grasses; they come from the Near East. They have common names, plenty of them, in Arabic, Turkish, Georgian, Hebrew, and Russian. For most of these languages the average reader does not even know the alphabet. *Aegilops,* with its initial *ae* diphthong, may look a little strange at first, but it is far easier for the average reader than its everyday equivalents in Turkish, or Arabic, or Russian.

This digression about *Aegilops* is worth the extra space if it interests you in the ancient and modern careers of this bristly little weed. One of the most brilliant pieces of current biological research has been the proof by two American scientists that

Aegilops played a leading role in the evolution of our modern bread wheats. From evidence too technical to discuss profitably here, a Texas wheat breeder named McFadden came to the conclusion that *Aegilops squarrosa* was the probable source of the DD chromosomes. He accordingly took emmer, one of our most primitive tetraploid (AABB) wheats, and crossed it with the diploid (DD) *Aegilops squarrosa.* He eventually produced, as one might have predicted, a sterile hybrid of the formula ABD. It had three sets of chromosomes, AB from emmer and D from *Aegilops.* Like most plants with three sets of chromosomes it was sterile, though otherwise it looked like a primitive bread wheat.

At this point a Missouri cytologist, Dr. Ernest Sears, joined in the work. He took McFadden's sterile hybrids and treated them with colchicine, a drug which if carefully regulated can prevent cells from dividing when their nuclei divide. With patience and good luck one may get a sector of a plant and eventually a whole plant which has developed from these affected cells and in which consequently the chromosome number has been doubled. By this technique Sears produced a hexaploid AABBDD from McFadden's sterile ABD triploids. As might have been predicted it was fertile and true-breeding. Furthermore, it was virtually identical with a primitive hexaploid wheat, known as spelt, which has been a minor European crop since Roman times. The spelt artificially produced by McFadden and Sears is fertile with European spelt and that the latter is a polyploid of *Aegilops squarrosa* and some simpler wheat may be taken as proved.

"Wait a minute," the reader may interject. "There were no scientists waiting around with a bottle of colchicine. How could the sterile hybrid double its chromosome number? It is one thing to show that something could have happened, and quite another to demonstrate that it actually did do so." This is an intelligent comment and it deserves a careful answer. In the first place, remember that in *Primula kewensis,* our original example of polyploidy, the fertile tetraploid branches arose of their own accord and on more than one occasion. In the second place, let me describe for you an Oriental grainfield. A typical one in modern Anatolia will serve since it is within the area where part of the evolution of wheat has taken place. The crop itself is a mixture of wheat and barley and rye. The wheats are a mixture, partly various tetraploids and partly hexaploids. In the field and growing around its margins are such related weed grasses as *Aegilops squarrosa,* Haynaldia, rye, and the like. Fields of this type are characteristic of primitive agriculture in the Old World and the New. Though the chance of a new polyploid arising in any one field is slight, with hundreds of thousands of such fields every year and several thousand years in the record, the elaboration of such a trigeneric hexaploid as wheat could almost be taken for granted.

Only one thoroughly familiar with the literature on the history of wheat can realize how completely the work of McFadden and Sears causes a realignment of the facts and theories in that complicated field. At the moment we are in exactly the same stage as are a group of people putting a jigsaw puzzle together when they discover that the pink pieces which they previously thought all belonged to some kind of a flower garden are partly to be put in the flower garden, but some of them belong in a sunset glow in another part of the picture puzzle. Spelt, for instance, has not yet been found in any ancient excavations, and it had previously been supposed to have originated from the bread wheats, to which it is obviously closely related. The synthesis of spelt from emmer and *Aegilops squarrosa* forces us to the hypothesis that it originated anciently and in the Caucasus or southwestern Asia where *Aegilops squarrosa* is native. It now

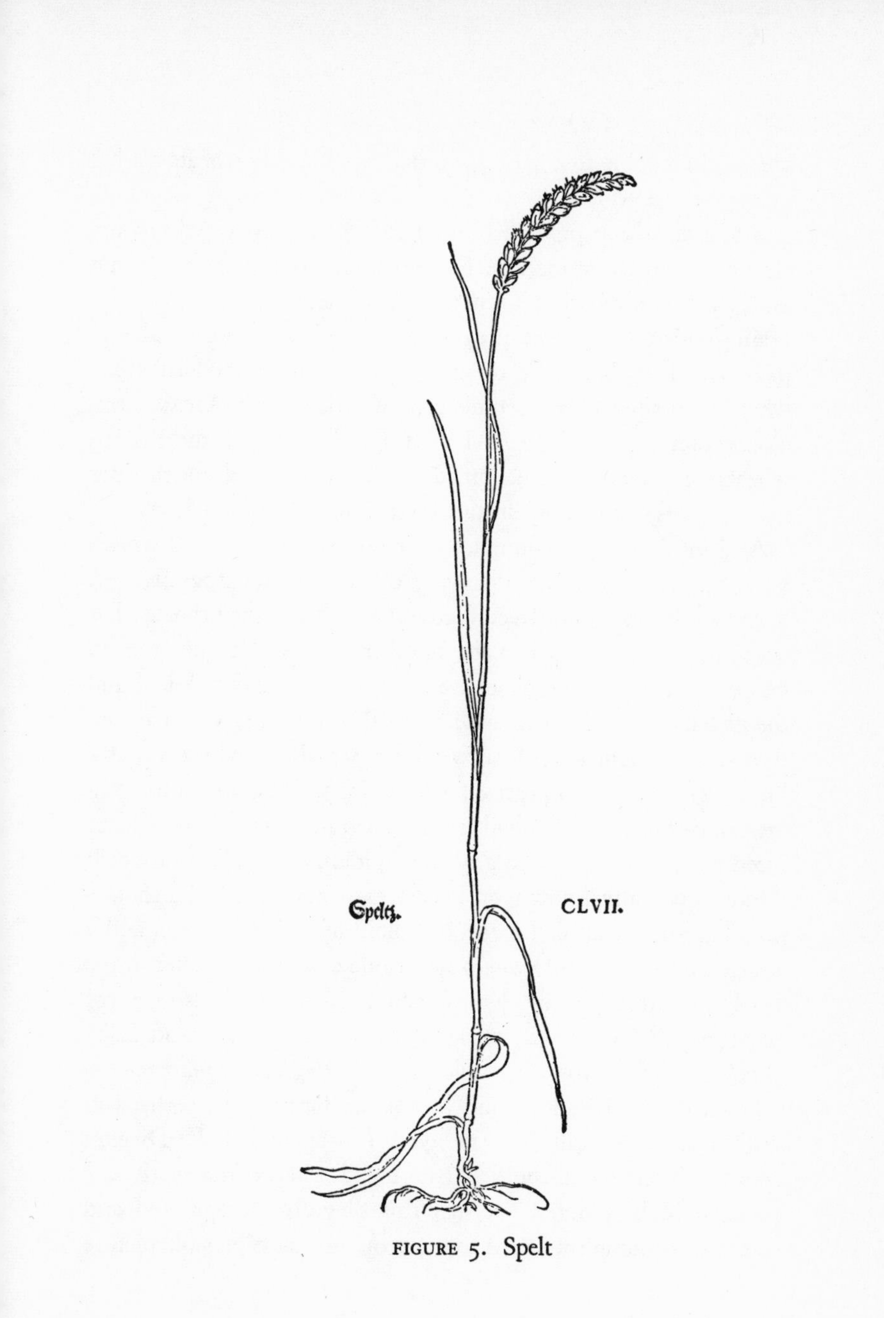

FIGURE 5. Spelt

seems as probable that spelt gave rise to the bread wheats as that they gave rise to it.

It had also been previously thought that the bread wheats are almost as old as emmer. McFadden points out that we do not know the chromosome numbers of the ancient Lake Dweller grains which have been previously identified as bread wheats. It is his view that since these are smaller than modern bread wheats and since they resemble certain little-known Asiatic tetraploids, they may well be AABB tetraploids allied to the Persian wheats and that the true bread wheats originated much later from crosses between spelt and these Lake Dweller wheats.

As a kind of visual summary of the origin of the bread wheats I have prepared the accompanying diagram of their family tree. It is a top-heavy sort of tree because the width of the branches has been made proportional to the number of sets of chromosomes; hence the tetraploid branches are twice as wide as the diploid and the hexaploid three times as wide. At the bottom of the chart are shown the original wild or weed grasses about which we still know very little. By a process about which we know next to nothing (and therefore left blank in the diagram) one of the wild or weed wheats eventually became the diploid Neolithic cereal which is now generally known by its German name, einkorn. Sometime later in Neolithic times a polyploid hybrid of einkorn and a weedy quack grass (*Agropyron*) became emmer. Still later in the Neolithic, probably by hybridization with weedy species of *Aegilops,* the tetraploid wheats such as the Persian and Lake Dweller wheats arose. Hexaploidy with *Aegilops squarrosa* (as confirmed by McFadden and Sears) produced spelt, perhaps in the Bronze Age. Still later the hybridization with Lake Dweller wheats differentiated the first true bread wheats from spelt.

It should be pointed out that this diagram, complicated and interwoven though it seems to be, is our simplest possible picture

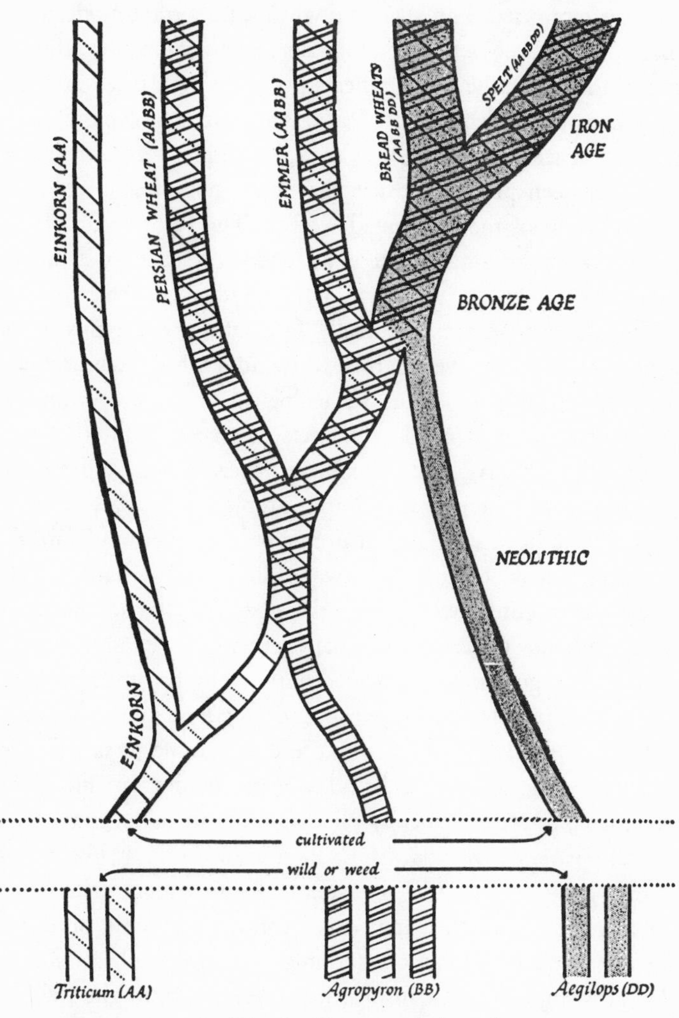

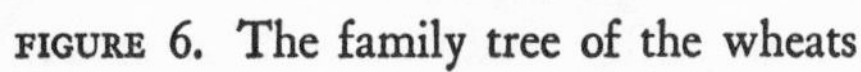

FIGURE 6. The family tree of the wheats

of the origin of the bread wheats; various other complications have been omitted. The macaroni wheats, important modern tetraploids, have been left out altogether, though cross-breeding between them and the bread wheats probably took place in ancient as it does in modern times. Various primitive polyploid wheats from the back corners of Asia, though of great technical interest, have not been put in, because they were not directly concerned in the origin of modern bread wheats. The full picture of these cereals when we finally put it all down on paper will make a much more intricate web than I have indicated here.

These minutiae are technically fascinating; were this a book about bread wheats we might well spend a whole technical chapter in discussing how the archaeological and the breeding evidence and the new cytological facts can best be fitted together coherently. This is no such book, however. What we are mainly concerned with here is the simple problem of how a few cytological facts have split an old problem wide open and forced us to consider a wide range of new possibilities. We had set out to study the origin of wheat, and the taxonomists had told us we were studying Triticum, and that there were such and such species in that genus. We now learn that our commonest wheats belong to the genera *Agropyron* and *Aegilops* quite as much as they do to the genus *Triticum*. Were it possible to sacrifice convenience to accuracy, our bread wheats could more fittingly be designated as *Aegilotriticopyron sativum* than as the currently accepted *Triticum sativum*. McFadden even thinks it likely that a fourth genus, *Haynaldia,* by simple hybridization without any polyploid realignments, may have contributed some of the distinctive characteristics of our tetraploid macaroni wheats. *Haynaldia,* growing as a weed in primitive grainfields, might well have crossed with some of the wheats. These hybrids, crossing back to the wheats again, would bring a little *Haynaldia* germ plasm into the wheats. In other words, we now realize that if we are to understand the wheats of the world, either as a key to prehistory or as the staff of life for a good portion of the human race, we must study a whole group of quack grasses and other weeds from the Near East and Asia. For the purpose of practical plant breeding the world now finds itself needing to know in detail about several humble weeds, plants so far removed from our ordinary lives that we have no common English names for them and must resort to such unwieldy technical vocables as *Agropyron, Haynaldia,* and *Aegilops* in discussing them!

Studies of the origin of rice are not as far advanced as those of the origin of wheat but there too the cytological evidence is making us revise our thinking. Most of the cultivated rices are diploids. One of the commonest of the weedy rices, *Oryza latifolia,* now widespread in both the Old World and the New, is a tetraploid. If at all closely related to cultivated rice, as has sometimes been supposed, it must have originated from rice, rather than rice from it. For rice, the cytological and genetical evidence has not yet advanced to the point where it can build up a theory of its own but enough has been done to indicate that theories concerning the origin of rice which do not take the cytological facts into account will not be valid enough to be of any economic importance in a rice-breeding program and will be of dubious value in interpreting the prehistory of the Orient.

It is in the sugar canes, wild, weed and cultivated, ancient and modern, that polyploidy has run wild. The cultivated sorts go all the way from hexaploids to twelve-ploids. It is already clear that such weeds as the Khas grass (*Saccharum spontaneum*) have played as important a role in the history of the sugar cane as they do in modern sugar cane breeding. Khas, sugar cane, and various wild and weed Oriental grasses form a gigantic polyploid complex. None of those experts now working with the problem would be brash enough to guess just which of these grasses represent the prehuman elements from which the sugar canes arose. Skilled scientists both in India and the United States have realized the scope any really fundamental investigation of this problem must take and detailed surveys and collections of all the wild and semiwild sugar canes of the world are now well under way.

For the bananas, cytological studies demonstrate that our important cultivated bananas are triploids, that they came from crosses between diploids and tetraploids, and that to proceed efficiently with the creation of new varieties we shall have to understand the classification and the cytology of wild and weed bananas quite as much as the cultivated kinds. We now realize that it is only by crossing a seedy (and hence wild or weedy) diploid with a seedy tetraploid that we can get a triploid (and hence seedless) cultivated variety. A triploid has three sets of chromosomes. The simplest way to get one is to cross a diploid which will contribute one set, with a tetraploid which will contribute the other two. Because the hybrid is a triploid it will be sterile and therefore the seeds will not develop. Every time we want a new variety of seedless banana we must go back to the original diploid and tetraploid stocks. In other words, with bananas, the cultivated varieties represent dead ends and to get new ones the breeder must go back to wild species or weeds.

To the history of no other crop does the cytologist bring more surprising clues than to that of cotton. The story of cotton is a complicated one at its very simplest. Hutchinson, the most recent monographer, reckons there are twenty species in the genus, four of which are known only in cultivation. The wild species are all shrubby, the largest being large treelike shrubs or small trees, the smallest wiry subshrubs. Each of the four cultivated species includes varieties that are more or less shrubby, the commercial crop being for all practical purposes an annual. The greatest complication in the history of cotton is that two of the cultivated species are native to the Old World and two to the New. Cotton has been grown in two widely separated parts of the world since very early times, exactly how early we do not yet know. In those brilliant early civilizations of the New World which preceded the Aztecs and the Incas, cotton was not only grown as a crop just as it was in the Old World; it was also spun expertly with the same kinds of spindles and woven on exactly the same sort of two-barred looms as in ancient India, but botanically it was not the same cotton.

In the Old World, the wide and expert use of cotton apparently began in the valley of the Indus, sometime before the third millennium B.C., though for all we know yet, the plant itself, and perhaps also the idea of using its fiber, may have come from Africa at a still earlier date. From these early centers cotton was slow in reaching the rest of the Old World. Herodotus did not find it in Babylonia or Egypt in 445 B.C., and it did not even reach China until about A.D. 700. In the Occident it was preceded by fabulous traveler's tales. It was described as a kind of "Vegetable Lamb" of the Orient, a true lamb, it was said, with a fleece that could be plucked and woven, but a lamb which was attached to a stalk and had roots which held it firmly anchored to the ground like any other plant.

When eventually the cottons of the world were brought together for study and scientific cotton breeding made its first beginnings, botanists and cotton breeders soon learned that the Old World cottons and the New World cottons were very different types of plants. Since there was apparently more than one species in each area and a host of wild plants and weeds more or less related to one set or the other, it was a perfect mare's-nest of a problem until cotton cytologists cut to the heart of it. There were, said they (and they had the facts to back up their contentions), two main kinds of diploids. One they called the AA type, the other the DD. The cottons of the Old World, wild and cultivated alike, were AA's. The wild cottons of the New World were mostly diploid and of the constitution DD. The New World cultivated cottons were all tetraploid, as were also some weedy cottons of the Caribbean and certain Pacific Islands, and they were all of the constitution AADD. In other words the New World cultivated cottons must somehow have originated from crosses between Old World AA's and New World DD's. How could this have happened? Well, up to this point the evidence is clear; beyond that we are in the field of inference and intuition, of hunches and speculation, of discussion, argumentation, and even, it must be confessed, of vituperation. But make no mistake about it, up to this jumping-off place the evidence is ironclad. For one thing, new tetraploids have been artificially synthesized by crossing Old World cultivated cottons with New World wild cottons to produce diploid AD hybrids, and then these have been converted into AADD tetraploids by doubling their chromosome number with colchicine.

One group of scientists believes that Asiatic diploids spread to the New World across the Pacific in Cretaceous or Tertiary times, and there in prehuman times gave rise to tetraploid hybrids with the wild New World cottons, then disappearing from the scene themselves, and leaving the AADD tetraploids as evidence of their trip. According to this view it was the wild AADD descendants of this ancient natural cross which were eventually domesticated in the New World. The rival explanation sees cultivated AA cottons brought to the New World across the Pacific by the same early wave of civilized peoples which are known to have come at least as far as Easter Island. According to this theory the AA diploids were grown in Peru where a DD diploid *Gossypium* is native, and comparatively abundant. Insects cross-pollinated the two and accidentally the new tetraploids arose.

And there the problem stands at present. The cytological evidence has cut to the center of the old confusion, produced a neat and verifiable theory, and then retaliated by throwing the whole problem into one of the most hotly debated topics in the field between anthropology and ethnobotany.

To what extent were there cultural contacts across the Pacific in pre-Columbian times? The archaeological and ethnobotanical evidence is not clear on this point. One set of anthropologists, the diffusionists, feel that there is strong evidence for the early spread of high culture from southeastern Asia across the Pacific to the New World. Another set of experts, equally eminent, maintain quite as stoutly that there could at the most have been no more than rare and sporadic contact, and that the cultures of the New World are matters of independent invention. There are able anthropologists both among the diffusionists and the inventionists. European anthropologists on the whole have been more in favor of transoceanic diffusion; American anthropologists, more likely to explain our indigenous high cultures as a flowering of the aboriginal American intellect. A generation ago, when leading American experts were more uniformly antidiffusionist than at present, a witty English anthropologist referred to this theory as "the Monroe Doctrine of American Anthropology."

A number of anthropologists from both camps have been quick to see that in the argument of diffusion versus invention, the evidence from cultivated plants is critical. You may independently invent a loom but you can't independently invent an Asiatic cotton. It would be more correct to say that the evidence from cultivated plants will be critical when our understanding of their classification is truly critical. The cotton breeders of the world, for practical reasons if for no other, have had to take an intense interest in the classification of all cottons, whether they seemed to be directly concerned with cultivated varieties or not. As a result, the classification of wild and cultivated cotton is more truly critical and on a more global basis than for any other world crop. When the beans, the squashes, the cereals, and the major pulses have been as carefully studied, then we shall not have to argue for and against trans-Pacific pre-Columbian contacts. There will then be clear evidence as to how strong the contact was, or how weak; how important or how unimportant, and even for how long it lasted. The evidence for the transfer of the sweet potato to Polynesia now satisfies all but a few die-hards, so evidently there must have been at least some slight contact. Discussing the evidence from cultivated plants when we know no more about most of them than we do now is just a waste of time and effort. As we have seen in a previous chapter the classification of nearly all cultivated plants is now in such a perilous state that the use of any such Latin term as *Phaseolus vulgaris* is just an elaborate and technical way of saying, "I do not know."

Since aside from the cotton evidence there are few really critical data on cultivated plants, when the cotton geneticists and cytologists arrived with brilliant new theories, they found themselves in the middle of a hot scientific battle, the lines all drawn and tempers a little ragged on both sides. All is fair in war, even in that most disgraceful of all wars, one between people who call themselves scientists. Consequently, the rival hypotheses as to the origin of polyploid Peruvian cotton, instead of being examined coolly and dispassionately on their own evidence, were grabbed or rejected by men with their minds already made up. The diffusionists hailed the hypothesis of cotton being brought to South America by man, here to produce New World cottons by hybridization with wild *raimondii* cottons; the inventionists damned it. Both sides were loath to consider it scientifically on its own merits. As a sample of inflamed scientific thinking, I present the concluding paragraphs from a recent review of Hutchinson's book by a leading American botanist who had already identified himself as strongly on the side of independent invention. After giving a bare outline of the cytological evidence and its interpretation he went on to say:

This conclusion will doubtless be highly acceptable to the diffusionists, as it represents diffusionism to the *n*th degree. . . . This is a beautiful illustration of basing a theory and drawing utterly untenable conclusions on the basis of one apparent or possible fact, essentially the genes of certain American forms of cotton. The authors absolutely ignore the historical aspects of the distribution of cultivated plants, wherein all trustworthy information is opposed to the theory they so easily "prove." One wonders had Lord Acton been a botanist or a geneticist, what his reactions might have been. It is a beautiful illustration of the pointedness of his apt saying, "The worst use of theory is to make man insensible to fact."

I can only conclude that the ideas of the extreme diffusionists, that is, those who must bring all or some of the cultural advances in America from the Old World via the Pacific route, are indeed "such stuff as dreams are made of." . . . Doubtless the diffusionists will consider me a botanical sceptic, but as a botanist I could only wish they would do some work in the fields of phytogeography and zoogeography and consider the field of early agriculture before sounding off too loudly on their pet theories. This is probably a forlorn hope, for as Jeremiah XIII:23 put the query: "Can the Ethiopian change his skin, or the leopard his spots?" We may expect, I suppose, more or less constant eruptions on the part of the diffusionist propagandists, as this or that individual sees, or imagines, developments in pre-Columbian America which simulate those of the Old World. Thus two mutually contradictory concepts may continue, the one based on preconceived theories and wishful thinking, the other based on the obvious facts that American civilizations were founded on a strictly American agriculture. It is admitted that the diffusionists have the inside track as to publicity, for often they come up with spectacular ideas. . . . As Liberty Hyde Bailey recently observed, "Our lives are guided by sophistries, half-truths, and inadequate judgments." This statement, to me, admirably sums up the situation that faces us in any consideration of the claims of the extreme diffusionists (and to this group I would now add certain theoretical geneticists) for to many of their expressed ideas, sophistries, half-truths, and particularly inadequate judgments definitely apply.

Though the above quotation is from an article by an eminent scientist in a reputable scientific journal, it does not illustrate a distinctively scientific approach to a complicated problem. The method of science is to get good, clear objective data which bear on a problem. When enough such data are at hand acrimonious critics have to put up or shut up, as they did with Pasteur. One who has such facts can make his point without sarcasm or the argument from authority. When scientists write picturesque polemics you may be sure of one thing. There are not yet enough good demonstrable facts in that field. It is the main business of a scientist to gather such significant data, but scientists are human beings, and it is often difficult, if not impossible, to keep one's loyalties and one's personal pride completely under the discipline of cool scientific reason.

These examples by no means exhaust the new evidence that has been turned up by the cytologists in the last few decades concerning the origins of cultivated plants. There are new facts and new insights into the origin of tobaccos, of white potatoes, of citrus fruits, of bluegrasses, of apples and pears, of roses, chrysanthemums, dahlias, and many other ornamentals, and of such modern domesticates as blueberries and strawberries. But quite as important as any of the specific information which it contributes to the problem of the origin of cotton or the origin of tobacco, is the bearing of the cytological evidence on the general problem of cultivated plants as a whole.

In the first place it gives us a wholly new appreciation of the importance of hybridization. After reviewing the cytological and breeding evidence, hybridization seems to have been a really major factor in establishing our crop plants. Fifty years ago any scientist who would have dared to suggest that our common wheats are quack grasses quite as much as they are wheats, or that seedless bananas could only be produced by crossing certain kinds of wild, seedy bananas, would not have been taken seriously. The modern plant breeder not only makes such suggestions, he goes ahead and proves them. The fortunate thing for our understanding is the way in which polyploidy preserves for centuries clear cytological evidence of certain hybridizations. The whole set of chromosomes is still there and with luck and perseverance it may be identified, even though the original cross took place in the Iron Age or earlier. Without polyploidy this clear evidence would have been lost. Those of us who are familiar with the cytological facts and have speculated about their over-all meaning, have suggested that hybridization was probably just as common among the diploids and may well have involved rare crosses with other genera. Such hypotheses can be proved for diploid crops but it takes much more time and requires more elaborate research programs.

In the second place the cytological evidence demonstrates the intimate connection between crop plants and weeds, an idea which we shall want to explore at some detail in a later chapter. Einkorn, our most primitive wheat, was essentially built up into a world crop by hybridizations with its own weeds, *Aegilops, Agropyron,* and *Haynaldia.* Our sugar canes came into being through a series of crosses between weedy grasses; modern sugar cane breeders are now using one of India's greatest pests, the Khas grass, in turning out improved varieties of cane. Weeds are evidently worthy of more serious study than they have received in the past.

Most important of all, the cytological evidence demonstrates that the history of any major crop is a long involved affair. It would be more precise to discuss the origin*s* of most cultivated plants instead of just their origin. For our bread wheats at least five important steps are involved; the evidence is pretty clear that these five steps took place not only at widely different times but in different parts of the world and among different peoples. For few of our crops is the story of domestication a simple one. It is instead a long complicated business, a fascinating and rewarding subject for study and experiment, but quite outside the scope of encyclopedic pontification. There was a time when anyone with a good general background in biology could look up the scientific names of our major crop plants (if he did not already know them), ascertain what parts of the world they were said to come from, and write a definitive article concerning the origin of agriculture. That time is past. One of the chief services of the new evidence from cytology and genetics has been to show up the experts.

V

The Clue from Diversity, or Science and the Bureaucrats

THE DICHOTOMY OF THE TITLE is deliberate. Throughout this chapter we must of necessity be concerned with two extremely diverse topics, the variation pattern of cultivated plants and the conspicuous and tragic story of the one great student of that subject. So great was the force of his intellect, so keen his imagination, that he brought an entirely new class of facts to the service of the subject of this book, the history of cultivated plants. So great was the force of his personality that he played a key role in world affairs. Had he been an obscure professor in some provincial university we might consider his theories apart from his life story. However, he was N. I. Vavilov, one of the leading biologists of his day; at the height of his power he had more people working on the history of cultivated plants than anyone before or since. This book being what it is, we cannot consider his theories entirely apart from his career.

The starting-off place for Vavilov's discoveries was that most unlikely subject, the study of homologous variation. This phenomenon had been noted and named before his time; he gave such facts a new significance. You yourself must have seen a good many of the facts, though you probably never puzzled over their interpretation. You have certainly seen a Lombardy poplar, the slender quick-growing tree which has a general outline very much like the upper part of an exclamation point. It is a freak form of the Italian black poplar, which turned up years ago. It is like the black poplar in everything except that its branches grow straight upwards alongside the main trunk instead of spreading out away from it. Cuttings were rooted from the original freak tree, and then cuttings were made from them, and cuttings from these in turn. So from hand to hand and from nursery to nursery these upright black poplars have spread around the world.

If you will go to any large arboretum, such as the Arnold Arboretum in Boston or the Morton Arboretum near Chicago, you will find that the Lombardy poplar is not the only exclamation point freak which has turned up. There is a similar form of the white poplar, there is an upright beech, an upright English oak; there are even upright sugar maples, spires of gold and orange when they turn color in the fall. As a matter of fact, nearly any tree will furnish us with an upright sport if we grow enough seedlings and keep a sharp eye out for such departures from the normal.

Now the law of homologous variation is that all such sports are found in plant after plant if we make a careful enough search for them. So we have the red-leaved varieties of our common trees, the red-leaved maple, the copper beech, the red-leaved plum, the red-leaved barberry, and even the red-leaved amaranth which was picked up by primitive man as a charm against evil spirits. So there are the weeping trees, and the dwarfs, and the snaky twisting types so repulsive that they are seldom seen nowadays outside of special collections.

As a young man Vavilov was much interested in cereal crops. Before World War I he went over to England to study with Sir Roland H. Biffen, the English wheat breeder. He had not been in Cambridge long before he found that for a young man with imagination there was more to be learned from Biffen's colleague,

Mr. William Bateson, a man who had taken for his province the study of variation, the evolutionary principle which Darwin's followers (but not Darwin!) mostly just took for granted as an underlying and ever-present force. Bateson encouraged young Vavilov to study homologous variation in cereals and other cultivated plants, since it seemed to interest him, and see what he could make out of it. He went to work enthusiastically. It was not long after this that still another of the Cambridge biologists met Vavilov on the street and quizzed him as to why he had changed professors. To him the young Russian was a promising wheat breeder, and here he was running off from Biffen, who was certainly turning out good wheats, to study with Bateson, whose ideas on variation were then too new to be fully appreciated so close at home. "Isn't Biffen a very good man to work with?" he asked Vavilov. "Ahh, yes. Beeffeen," said the latter, trying to make his ideas carry through local prejudice, struggling with his imperfect knowledge of English and a thick, Continental accent. "Beeffeen, yess, he is a gude man, a fery gude man, but you see, he has no phee-low-so-*phee*," with a dynamic accent on the last syllable of the last word.

The first thing which Vavilov discovered was that the law of homologous variation really worked for the cereals. If you found a freak or primitive form of one cereal, it could nearly always be matched in the others if you looked long enough and in enough places. If there was a black wheat, then you could expect black oats, and black barleys. If there were winter and spring varieties of one cereal, then winter and spring varieties could be found for the others. These facts made it possible to set up a system for classifying the myriad varieties of cultivated cereal crops. By listing all these peculiarities one produced a filing system with scores of pigeonholes; if a pigeonhole was vacant one had only to look and see if other cereals had something in that niche; if they did,

Vavilov would predict that it must be there for this crop too, and he was usually right.

From these elementary matters he went on to a most amazing discovery. There was a geographical regularity to the distribution of these freak forms of the different cereals. If Anatolia in central Turkey was found to have a large number of peculiar types for one crop, one could predict that it would be a center of diversity for several other crops as well. More than that, one could predict the kind of variant which would turn up in such centers of diversity: it would usually be the primitive type most like wild grasses. Wild grasses are rough and hairy, their seeds scatter and fall soon after they are ripe (shatter is the technical term), they are frequently more or less dark-colored. Sophisticated modern varieties are nonshattering, they are not unpleasantly bristly, and they are light-colored enough to make a white meal when they are ground. At all his centers of diversity Vavilov found a concentration of primitive variants, shattering, bristly, and dark-colored. When he turned to other cultivated plants than the cereals, he again found primitive types in these same centers.

By now the Russian Revolution had taken place and Vavilov, naturally a democratic, forward-looking sort of person, had become a power in Russian agriculture and Russian biology. Under his dynamic leadership, expeditions were sent out to all parts of the world to locate these centers of diversity and to bring back breeding material for the hungry Soviets. In a few years these expeditions laid the foundations for our modern understanding of the world's centers of primitive agriculture. Vavilov and his associates established the fact, and it is a fact, that the great bulk of variation in our crop plants, and to some extent in our weeds, is concentrated in a few small areas. These are: (1) Southwestern Asia, with the heaviest concentrations in Kashmir, Turkestan, Persia, and the mountains of Anatolia; (2) India and the Malay

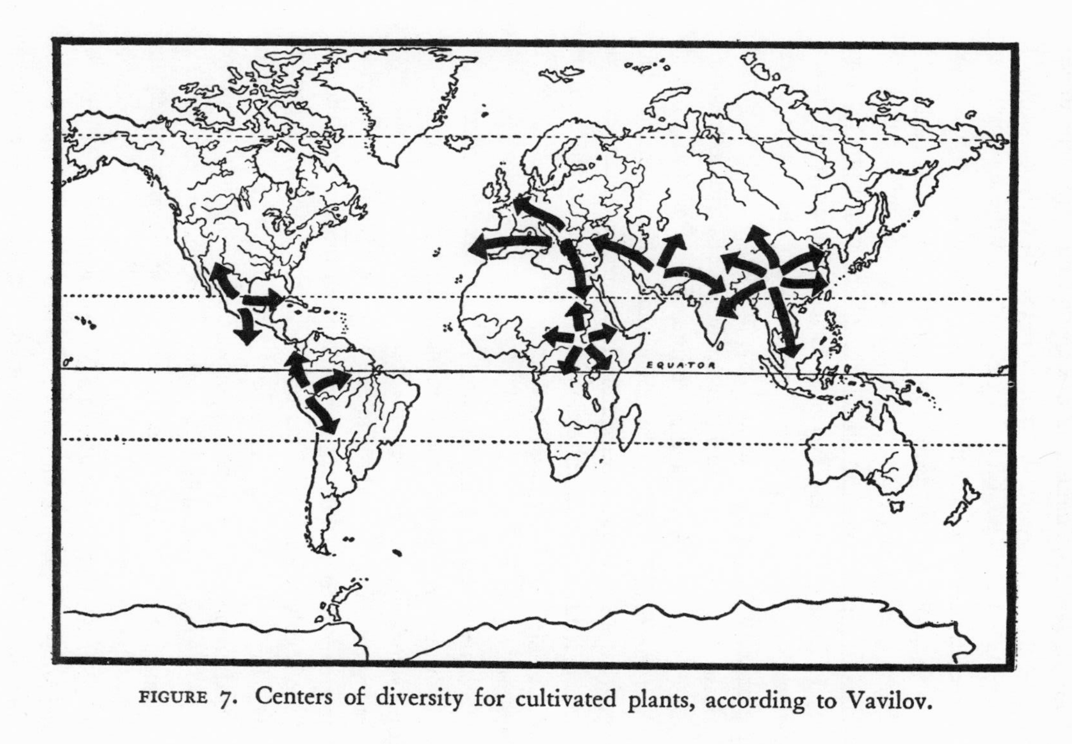

FIGURE 7. Centers of diversity for cultivated plants, according to Vavilov.

Peninsula; (3) The mountains of Central China; (4) The mountainous border of the Mediterranean—Palestine, Syria, Italy, and Spain; (5) The New World, with centers in the mountains of Mexico, Guatemala, Colombia, Peru, and Bolivia, and another center in eastern South America. In these few centers, an insignificant fraction of the earth's surface, is much more variability of the crop plants than in all the rest of the world put together.

Since these centers are known to have been the sites of ancient civilizations, Vavilov's first conclusion was that these centers were the places where agriculture had first developed and that in part, at least, they were the actual places where crops had been domesticated. Later, as he worked more intensively over such crops as wheat, he realized that some might quite as well be secondary centers of origin where something happened to an ancient crop to send it off in a new direction as when, for instance, emmer wheat united with *Aegilops* to form spelt wheat, as described in the previous chapter.

In his very last communications Vavilov approached the idea that these were centers of ancient agriculture, but not necessarily the even earlier places where the individual crops had undergone domestication; but by this time he was increasingly occupied by far graver matters. His entire program fell into disfavor, he was semiofficially tried before his fellow scientists for adherence to ideas incompatible with the official Marxian philosophy, and he ended his days in disgrace, supposedly in a labor camp in Siberia.

The story of his trial and the outlawing of conventional genetics in Russia has been reviewed in detail in books, magazines, and editorials. If you want to get as straight an account as possible, read *Death of a Science in Russia,* since it is made up of the original Soviet documents and accounts, with a minimum of editorial comment. What has not been stressed, what has only been hinted at, is the much more serious fact that Vavilov's problem in Russia is merely in an intensified form the problem which scientists face in this country and in all countries, an increasing struggle between them and the bureaucrats. In this country the bureaucrats are powerful, and more than one scientist has been persecuted by them, as for instance the distinguished director of the Bureau of Standards. After the Russian Expedition to Latin America had discovered the polyploid complex of wild and semiwild potatoes, an alert administrator in our Department of Agriculture proposed to send off a plant explorer to bring back a collection for American potato breeders. His budget was held up to public scorn in Congress because it contained the word *polyploid*. An ignorant legislator made public sport of the word and damned the whole project, which was of great potential practical importance, just as was Vavilov's. In Russia the bureaucrats are all-powerful; a Dr. Condon cannot retire and go into industry; a professor cannot without recanting seek sanctuary in a privately endowed university. The Soviets were the first great power to take biology seriously and to paint biologists as noble servants of the state. In this country biologists have a greater freedom, partly because their importance is not yet generally recognized. The man in the street still tends to think of botanists as odd little persons who putter with plants, and of entomologists as funny-looking creatures who caper over meadows with butterfly nets. When he finally realizes that biologists are working not only with the sources of his daily bread, but with his sex life, his ways of thought, his possible evolution, his most efficient extinction, then biologists will be even more hedged with throttling red tape than are modern physicists.

One of the great problems of the modern world is learning how to advance science without stifling it. Eager young men with bold imaginations are frequently wrong, but they are sometimes magnificently right, even in the face of the united disapproval of authorities in their own field. When Dr. East a half century ago came along with the idea of hybrid corn, the maize authorities in the experiment stations and the Department of Agriculture were against this wild idea. Had we had a centralized scientific administration his new notion would have died right there. Fortunately, a tolerant director shipped him off to a small research station well outside the corn belt with the statement that if there was anything in his theories, the new job would give him a chance to prove it and if, as seemed likely, he was all wrong, he would be working outside the corn belt and it would do no harm. Even after he and others had piled up good solid data to show that there was something behind his idea, the experiment station where he began his work remained aloof and was the last of those in the central corn belt to take an active interest in the new method of breeding corn. The kind of administrator who can survive in a modern experiment station is seldom the kind of person to welcome novel approaches to old problems. What the world needs if science is to forge ahead is some kind of system which will allow the bold young men to try out their wild ideas in spite of the judgment of older and usually wiser men. Dr. Henry Allen Moe, the distinguished Secretary-General of the Guggenheim Foundation, has recently pointed out forcefully that the encouraging of these new and fruitful ideas is a most delicate matter. He points out specifically that great foundations, private or governmental, set up for the encouragement of science, may with the best of intentions throttle scientific progress. If, using the best brains available, they draw up programs of scientific advancement, rather than choosing brilliant leaders and letting them find their own way, science may be prevented from taking the strange new direction which would have been most productive.

It is the old, old contest between the prophet and the pharisee. A stable society needs both, the prophet to find the new road, and the pharisee to keep us on the track after the road has been marked out.

The lesson for us from the Vavilov affair is that such clashes between science and bureaucracy are inevitable under any system of government. The same thing happens in this country as in Russia, but on a much milder scale. It is tempered by our system of checks and balances. In the last few decades, however, as the importance of science has become more apparent, it has gotten more and more under bureaucratic control. Our enormous government research laboratories tend to be staffed by able plodders who do not question well-established concepts. What we need for our national survival are laboratories in which able scientists can work unhampered by Congressional investigations, yellow journalism, cheap politicians and directors with established reputations — protected at times even from the supposed experts in their own fields.

As for Vavilov's centers of diversity, we know little more about them today than when his last paper appeared. His ideas and techniques were so far from the main currents of biological research in this country and in western Europe that although his work attracted a great deal of attention, no other laboratory or investigator was stimulated to work along similar lines. The evidence for his centers of diversity of agricultural crops is convincing; his interpretations of this phenomenon are more doubtful. In part, at least, the persistence of primitive varieties in mountainous regions is due to the generally innate conservatism of hill peoples. The American hillbilly with his old-fashioned dress and outmoded speech has his counterpart in hilly and mountainous country in other parts of the world. One of our American barley experts once pointed out that a strict application of Vavilov's principles would make eastern Tennessee the center of origin for the barleys of the New World.

One of the few people to give Vavilov's facts and hypotheses the consideration they merit is Dr. Jack Harlan who went to Anatolia, one of Vavilov's ancient centers, as a plant explorer for the United States Department of Agriculture. He described the mixtures of crops and weed crops and weeds in the grainfields of this region and raised the question as to how much the agricultural techniques of the region had to do with keeping up its crop diversity. His descriptions of Turkish fields remind me on the one hand of the mingled fruit and vegetable gardens of Guatemalan Indians and on the other of the fields I saw in northern India on the edge of another of Vavilov's ancient centers. We Europeans and American Europeans are so accustomed to thinking of agriculture in terms of fields all of one kind of plant, of one variety of wheat stretching as far as the eye can reach, of one hundred acres of maize, all of one variety and virtually weed-free, that we need some such contact as this to bring us back to what agriculture must have been like through its long Asiatic, African, and Mediterranean beginnings.

In India there is seldom any such thing as a field all of one variety of wheat; usually there is not even a field which is all wheat. The interplanting of various crops is the rule, not the rare exception, and frequently actual mixtures are planted and harvested and ground and used together. The grainfields are commonly interplanted with rows of oil mustards every three feet or so. The mustards are themselves mixtures in which two species of Brassica (belonging to three varieties) and one species of Eruca take part. The grain is more frequently mixed than not. It is usually a mixture of wheat and barley, and there are different local names for a crop which is all wheat, for one which is mixed but predominantly wheat, for one which is about half-and-half, and for one which is more barley than wheat. These are by no means the extreme in mixed cropping. Walking through the fields near Allahabad I saw a grainfield of mixed barley and wheat in which there were blooming the purple flowers of the field pea and the little red flowers of the small chick-pea. There were a few mustards, whether by accident or design I cannot say, and some plants of one of India's minor pulses, *Lathyrus sativus,* a field relative of our sweet pea.

I chanced to see in a few weeks that such mixtures of grains and pulses with a few mustards are commonplace in various parts of India. The wheat experts with whom I worked there assured me that as one got back into the lower slopes of the Himalayas, the grainfields became even more complex; mixtures of the various species, races, and strains of wheat were so commonly interplanted that wheat varieties in our understanding of that term scarcely existed. They are so variously mixed that methods of sampling and charting native fields will have to be worked out before we can even discuss in a general (but accurate) fashion what kinds of wheat are grown in the Himalayan borderlands. If agriculture did begin with dumpheap-orchard gardens, as will be suggested in a later chapter, then these conservative areas are growing crops under conditions more closely resembling those of Neolithic times. From this point of view Vavilov's ancient centers may perhaps be centers of survival rather than centers of origin. Much the same idea has been expressed by Harlan in his description of the Anatolian grainfields. From this point of view, Anatolia, Turkestan, Abyssinia, and Transcaucasia represent out-of-the-way corners in which the persistence of ancient agricultural methods has allowed ancient crop diversities to survive.

At one time Vavilov revived an old idea of William Bateson's to explain the evolution of crop plants. It was certain that they were most like their wild ancestors at these centers of diversity and became progressively more and more different as one proceeded towards the margins of their culture. Could it be that evolution, as Bateson had once suggested, was some kind of an unpacking process, that by progressive losses (as, for instance, loss of the ability to make color, loss of heavy bristles, etc.) the evolution of crop plants went on? Vavilov carried this suggestion even farther. His official duties carried him to many parts of the world and he was a keen observer. He noted that the native floras were markedly more variable in some parts of the world than in others; did they too have their ancient centers of diversity, and was all evolution a matter of a gradual degeneration from this primitive condition? He raised the question, but so far as I am aware, never did any work in this field. I have personally been much interested in his suggestion because for twenty-five years I have been working on techniques to study just such questions. It is only in the last few years that they have advanced to the point where I could use them to get evidence for venturing a preliminary opinion. From the relatively few plants which I have studied intensively, I would suppose that these areas of greater variability are points at which floras (and presumably faunas) which had previously been separated have come together and hybridized. The clearest example in the eastern United States is in northern Florida. Southern Florida was once, and perhaps repeatedly, an island or a series of islands. When it was eventually united to the mainland two related but different floras must have met in what is now northern Florida. We should therefore expect, on the hybridization theory, to find it today to be populated by species, some of which are excessively variable. We already know, for about a dozen genera, that this is true; eventually we should have enough data either to establish the hypothesis firmly or to set up a better one.

My experiences in this field have made me wonder if something comparable may not be one of the reasons for the extreme and localized diversity in some of Vavilov's centers. Suppose that very early in the origin of agriculture, say in the Neolithic or earlier, an agricultural people had colonized India from Africa, taking with them the seeds to start one of their crude, weedy garden-dumpheap-orchards. When they reached India some of their crops would then be in an area where wild relatives were growing, different from any of those they had previously been mingled with. The resulting hybridizations would have given a special variability to this area which might have persisted for a long time. The same sort of process would have gone on, as we shall describe in a later chapter, when the Great Plains sunflower reached California and hybridized with the little endemic sunflowers of the serpentine areas.

I have been able to test this theory in a tentative sort of way in one of Vavilov's major centers, highland Guatemala. There, in an area only a few hundred square miles in extent, maize is as variable as in all of the United States put together, with part of Mexico thrown in for good measure. I was able to work in this area for a few weeks, a year or so after my return from Mexico where I had put in seven months learning how to measure effectively the variation in native cornfields. From my preliminary results in Guatemala it would seem likely that much of the variation in Guatemalan maize comes from a mingling of Guatemalan and Mexican types. Below Quezaltenango there is a broad valley which leads up from the seacoast in among the high volcanoes of the mountains. It is a natural invasion route. It was here that the Spanish marched in following the same general path as the invading Mexicans of much earlier pre-Columbian times. This is just the part of Guatemala in which maize is the most variable, in which there are more kinds of varieties in one town and more differences from one town to another. May it not be that one of the interesting chapters in the history of maize occurred when raiding Mexicans brought new types of maize down into Guatemalan territory and set off the mingling, not merely of two varieties of maize, but of two distinct races? The hypothesis fits such facts as we have and runs counter to none of them. When a thorough study has been made of variation in Guatemalan maize, when similar studies have been made of a few other crop plants, then we shall be ready to sit down and assess the forces which built up one of the centers discovered by Vavilov and his associates. As Jack Harlan suggested, these treasuries of potentially useful variability deserve careful study.

VI

How to Measure an Avocado

IT IS THE SIMPLEST QUESTIONS which are the hardest to answer. What is life? How are time and space connected? Such common, everyday matters are among the chief concerns of our greatest scientists and philosophers. One of these simplest of simple questions has been hiding around the corner from us ever since the first pages of this book. What is a wild plant? Why, a plant that is growing wild, to be sure. But how does one know when it is growing wild? Here on the Missouri terrace where I am writing it is easy to answer the question by pointing to examples. The day lilies, bugle vine, and honeysuckle along the terrace are cultivated plants, the corn across the fence was planted by our farmer neighbor this spring, the sycamores and silver maples beyond it are wild—the very oldest of them were growing there before the white man settled in the Meramec Valley.

In other parts of the world where man has lived for a longer time, particularly in the rainy tropics where the lush rain forest reclothes the landscape almost overnight, this simple question may become so difficult that no one can answer it precisely. Paul Standley once published a technical flora of the Lancitilla Valley in Honduras with an introduction so perceptive that it should be generally read by those who travel in the tropics. He wrote:

The forest of these hills has every evidence of being perfectly primeval. There are all the marks that are supposed to furnish reliable criteria upon this subject—giant forest trees in great variety, an abundance of corozos and other tall palms, and a great profusion of the more significant small palm species, tree ferns, and many other plants that never are known to exist in second growth forest.

But what is primeval forest in Central America? Who knows? Upon the exposed slopes high up in this forest one sees abundant shards, pieces of clay vessels now so soaked by the perennial rains that they may be crumbled like chalk between the fingers. Is it unreasonable to suppose that hundreds of years ago these hills may have been cleared and planted with corn, just as they are being cleared today by the descendants of those aborigines? If these transient clearings are surrounded by virgin forest, will not the native plants at some time, after the clearings have been occupied by *guamil* reseed them with forest species? Is it not possible that these cacao bushes and sapote and avocado trees are remnants of plantations of long ago?

When in such a jungle we encounter a small-fruited relative of one of our cultivated fruits, what is its status as a wild plant? Who knows? It may be a remnant of a little orchard deliberately planted at a town edge and persisting in the engulfing vegetation. It might be a self-sown seedling from such a tree. It may have started unasked and uncared for on a refuse heap before the town was abandoned from a fruit brought either from a distant market or from a nearby woodland. It may be as genuinely wild as any part of the native vegetation. There is great variation among the wild-growing avocados (notice that I did not say the wild avocados) of Central America. There is even greater variation among the cultivated sorts. How much of the variation among the wild-growing ones is due to the connection with previous cycles of cultivation? We cannot yet answer this question, but as I shall demonstrate in a moment there are now methods available which will help us to an informed opinion; a very small beginning has even been made at applying them to this particular problem. Before going into these details I want to assure you that this is not just an avocado problem nor is it confined merely to a few tropical plants. It is one of the basic difficulties in getting a really critical understanding of what happens when plants are domesticated. There are no easy answers to these difficulties but if we know that they exist we shall be properly critical of those who sound off with an air of authority, nor shall we fool ourselves by accepting the easiest answers.

The red and yellow mombins (or "hog plums") are even more of a problem than the avocados. It is a little hard to talk about mombins because they have so many different names, none of which are really widely used. They are popular fruits in various parts of the tropics, and are about the size of plums. In Latin America they are frequently called the "ciruela," that being the common Spanish name for the European plum. Botanically they go in the genus *Spondias,* and are not at all closely related to plums, though like that fruit they are more or less acid and have a large central stone. Some are red, some are yellow, some red with a flush of yellow. Most Europeans do not consider them much of a delicacy but to the Indian populations of Central America they are one of the pleasures of life. One sees baskets and piles of them in every native market. Indians coming into town with bundles of produce on their heads munch them as they hurry along at a half trot, and when they are in season towns like Antigua have a superficial paving of mombin pits on top of the old cobblestones. How many species of mombin are there, and where do they come from? The books will give you a simple answer and so will those authorities with nothing better than book knowledge at their command. Ask Dr. Louis Williams of the Escuela Agrícola Panamericana in Honduras, and you will get no such simple answer. He has had the full-time job of collecting and studying economic plants in Latin America for over a decade. He throws up his hands and says, "Who knows? There may be two kinds; there may be fifty. The Indians have been gathering them and spitting them out again for we don't know how long. They have been planting selected forms in their native gardens and these native gardens have reverted to woodland. How are we ever going to know where they all came from?"

Such things happen in the Temperate Zone but on a gentler scale. The old pastures of New England are so full of seedling apple trees that Wallace Nutting made himself a good business going back and forth across New England in the springtime, photographing the wild applebloom and selling tinted photographs on a nationwide scale. Few of these old trees were deliberately planted by man; boys and squirrels had something to do with many of them. At any rate all of them are Old World apples; none of them were here when the first European colonists arrived. Farther west and south where our American prairie crab apples are native, things are not always so simple. Here one sometimes comes upon a "Soulard Crab," a rampantly growing hybrid of the domestic apple and the prairie crab. It is a strange sort of plant, its bloom midway between the dainty pink and white of the cultivated apple and the exotic salmony pink of the prairie crab. These Soulard crab apples are a demonstration as clear as a laboratory experiment of the tangled histories man has brought to the apples of the world. Our native prairie crab apples and the domestic apple (whether man-selected or self-sown) are so strikingly different that just by careful comparisons one would be pretty certain that the Soulards are hybrids, but we have still better proof. In the last half century the Western world has taken an increasing interest in the crab apple as an ornamental plant, and similar crosses have been made artificially. Imagine what happens when cultivated apples run wild in Asia, the part of the world they probably came from in the first place and where there are many different kinds of native apples. In New England it is clear to see that the wild apples in the pastures came from the cultivated crop. In parts of Asia where various species of apple are native, who is to know (on our present evidence) which of the wild-growing trees are ancestral to those of the orchards and which have sprouted from an apple picked in a village garden? Man has been in Asia a long time; we know from actual excavations that he has been munching apples since the Neolithic and probably very much longer. When one of our plant collectors from the United States Department of Agriculture brings back a wild-growing apple from Turkestan or the Caucasus or eastern Turkey, he may be bringing a plant some of whose ancestors were first carried from their native habitats over a hundred thousand years ago, some of whom were actual orchard trees within the last five or ten thousand years, so that they had journeyed from one part of Asia to another. In terms of prehuman botany a little wild-growing seedling from the Caucasus may unite the blood of a Turkestan and a Himalayan species with one from the Caucasus, and be quite as much a mongrel as the average American citizen.

It was when I was working with spiderworts that I gradually evolved a method for studying such problems as these and eventually I had the chance to try them out on avocados. By the time Dr. Woodson and I had finished monographing the American species of Tradescantia, it was evident to both of us that hybridization between the various species might well be of evolutionary importance in that group of plants. In those days there was a good deal of argument going on about this subject, much of it merely the voicing of opinions by biologists who had been studying other matters but had had some opportunity to observe hybridization or the lack of it. It seemed to me that instead of just arguing it would be better to set up some kind of a method for measuring the effects of hybridization. Postponing the fascinating question of whether or not hybridization was generally important in evolution as a whole, I began to work out ways for measuring what it was and was not doing in our spiderworts. I will not bore you with the steps by which there were gradually evolved a few simple methods for studying such matters, nor cite in detail the various groups of plants, spiderworts, irises, asters, loco weeds, which have been studied in detail by my students and collaborators. Eventually a simple and generally applicable technique was worked out, one which I finally used on Mexican avocados, as I shall tell you in detail a little later. One of my assistants always referred to it as "whiskers" for reasons which will be evident when we get to studying the next figure; and that term is still used informally around the laboratory. However, such sprightliness is not considered good form among scientists so when it was finally formally published it was described under the mouth-filling phrase of "pictorialized scatter diagrams."

To show you how very simple a technique it really is, let us invent two species, one with a red flower and one with a white, and demonstrate how it would be used in studying their hybrids. We had better make the two species sound a little more respectable by giving them Latin names, so one shall be "rubra" and the other "alba." Let us make them good species, differing in a whole set of characters as good species always should. Alba shall be low, narrow-leaved, and hairless; rubra tall, broad-leaved, and hairy. Alba's petals shall be quite simple in outline; rubra's shall have a deep notch in the apex.

Now let us go and measure exactly a whole bunch of rubras from a June meadow, and a similar collection of albas from a nearby woodland. For each plant we must take the height of the plant, the width of an average leaf, the color of the flower, some precise record of the hairiness, and the depth of the notch in the petal. If we are going to keep our eyes on all five of these things at once, we shall have to cook up some homemade expedient if the method is to be kept simple. Statisticians can do a pretty good

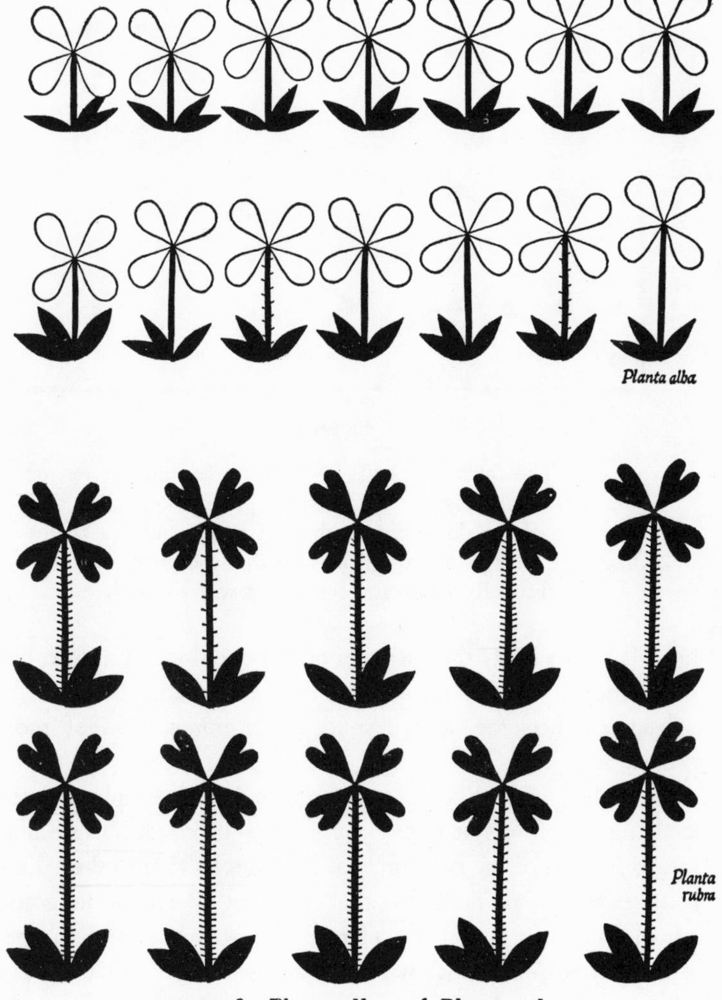

FIGURE 8. *Planta alba* and *Planta rubra*

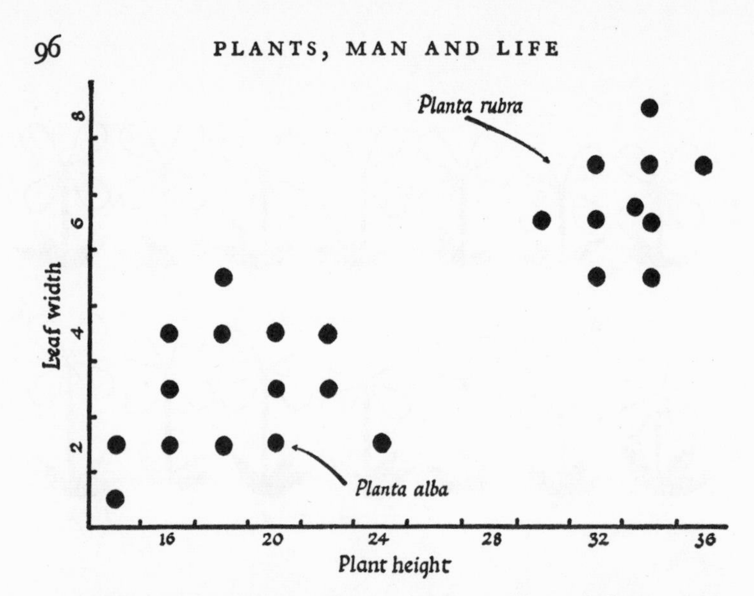

FIGURE 9. The fourteen plants of *Planta alba* and the ten of *Planta rubra* from Figure 8. Each dot represents one plant and the position of the dot indicates the leaf width and the height of that particular plant. It will be seen that the plants of rubra are distinctly taller and somewhat wider-leaved.

job when they study the variation of any one thing at a time, but when they study five all at once the mathematics becomes involved, cumbersome, and beyond the comprehension of most people.

Now the good philosopher Descartes had a simple way of studying two forces at once by using a squared-off surface and measuring one of the two forces at right angles to the other. This is an easy enough dodge so that it might be taught in grade school, though it is usually deferred to high school and college. It is really so simple that we use it all the time in our everyday affairs in the modern world. On stock market curves, for instance, we let height above the base line represent values in dol-

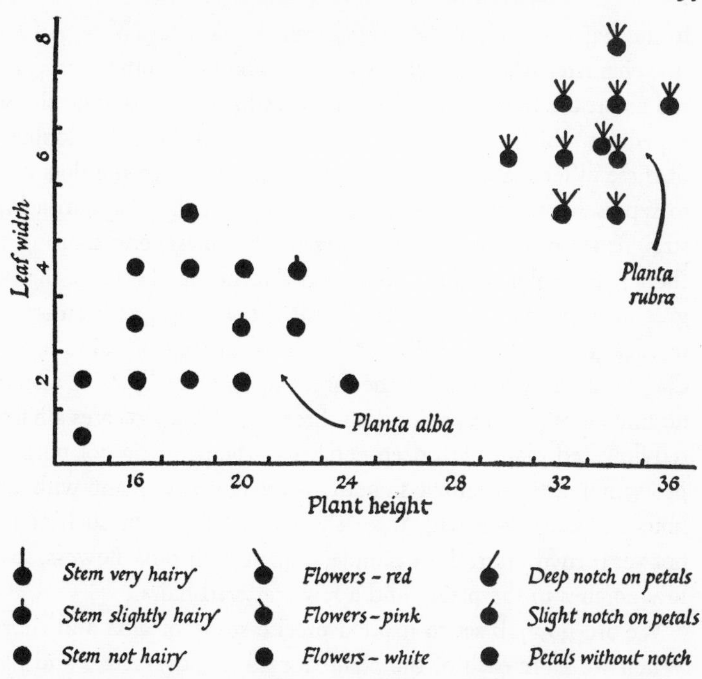

FIGURE 10. The collections of *Planta alba* and *Planta rubra* diagrammed just as in the previous figure except that indications as to flower color, hairiness of the stem, and notching of the petals have been added to each dot.

lars and cents, while distance along the line represents time in weeks, months, and years. Using Descartes's method in another simple form we may measure plant height along the bottom and top of a diagram and leaf width along the sides. Taking our collection of ten plants of rubra and fourteen of alba we can diagram each one with a dot. The speckly figure which results is a precise way of showing that we had ten tall plants with wide leaves, and fourteen short ones with narrow leaves. Each round dot represents one plant, the entire diagram shows us just how tall and how wide in the leaf was each of the twenty-four. It takes care of two of our five characteristics but tells us nothing about the other three, the red color, the hairiness, and the notches in the petals. The simple expedient which I finally hit upon was to record the development of these other features by little lines radiating from the dots used to represent individual plants. For this particular diagram a line straight up on the page designates the hairiness, one slanting to the left the color of the corolla, and slanting to the right the degree of notching. It will be seen that the scoring has been set up in such a way that the distinctive characteristics of rubra are all diagrammed by long lines, the three opposite attributes of alba by no lines at all. Hence, a dot with three long lines indicates a hairy, red-flowered plant with deep notches in the petals; a dot with no projecting lines means a smooth, white-flowered plant with unnotched petals; one with three short lines is a plant such as has not yet turned up in this example, a plant with pink flowers, shallow notches in the petals, and a few scattered hairs.

We are now all set to make a precise study of alba and rubra. When we score each of the plants for all five characters and plot the resulting diagram, we find they were not quite as uniform as we had thought. There are a few plants of intermediate pubescence and with leaf widths which are intermediate. The two albas with intermediate pubescence are among the largest of that species; the one rubra with intermediate pubescence is among the smallest rubras and furthermore it is also one of the two rubras with narrowest leaves. It looks as if the combination of characters which distinguished alba from rubra is not quite as sharply defined as we had previously supposed. There does seem to have been a very slight mixing going on between these two species.

After we have gone this far with the problem we chance to come upon a cutover woodland (now, all of these details are as made out of moonshine as *Alice in Wonderland,* but all the imagined facts are paralleled by real details on real plants in real woodlands which had been cut over) which has albas growing in it. Some of them are very like the albas we have seen elsewhere. As a whole, however, they are much more variable. There are a few plants rather like queer-looking rubras and a few neither exactly alba or rubra, though reminding us of both. Plotting all this collection by the same scale we had developed previously produces the final diagram. By studying it and comparing each plant with the little glyph which represents it on the pictorialized scatter diagram, we can eventually get the whole complex problem summarized. These plants are albas, varying in the direction of rubra. Some resemble rubra in one way, some in another; a few of them resemble rubra in many different ways. Though it is not inevitable that if a plant has somewhat the color of rubra it must also be larger as is rubra, there is a very strong tendency of this sort. It is, in other words, a population of alba plants contaminated by previous hybridization of some of their ancestors with rubra. The exotic characters which came in from rubra tend to stick together on the average. There is a loose but not inevitable association of the reddish, most notched flowers with the hairiest, tallest stems and broadest leaves.

After some practice in the use of this method it is possible to extend its use to mongrel populations in which only one of the parental species is present. In our diagram, for instance, there are no full-blooded rubras. It would not take much experience with such methods to predict that the variation in these plants from the cutover woodland was most likely the result of pure alba hybridizing with a species which had red flowers, long stems, broad leaves, hairy stems, and deeply notched petals. With even more experience it has been possible for those trained in this method to draw up descriptions of the second species even when they have never seen it, descriptions so detailed that the hypothetical source of the

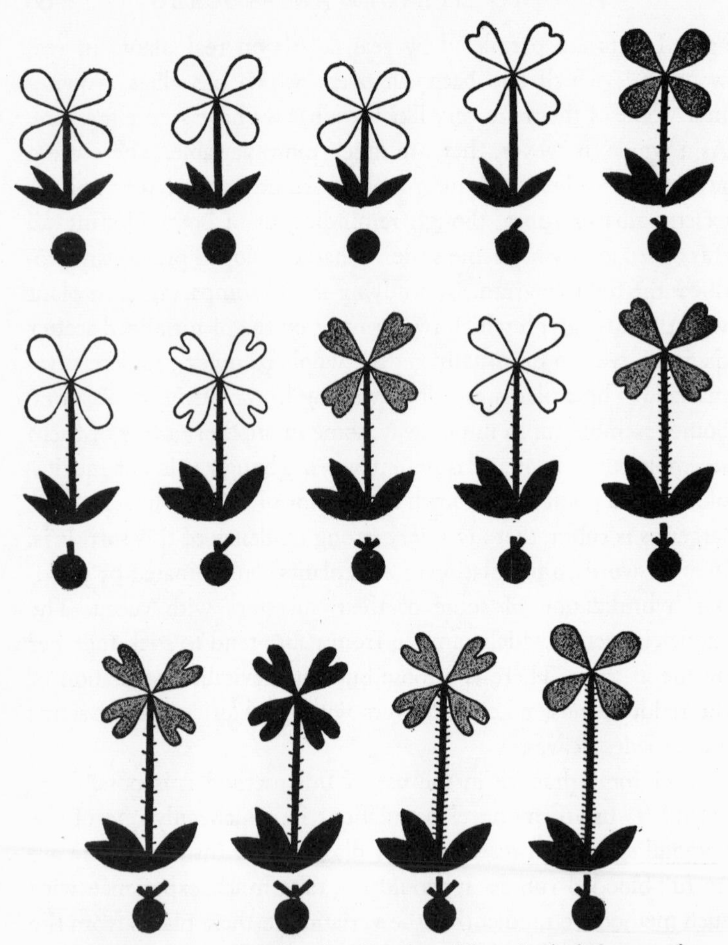

FIGURE 11. A mongrelized collection. The glyph below each plant shows how flower color, hairiness of stem, and notching of petals are represented. The arm to the right represents the depth of the notch, that to the left the color of the flower; the center arm shows how hairy the stem is. For instance, a dot with no arms represents a plant whose flowers are white, without notched petals, and whose stem is not hairy.

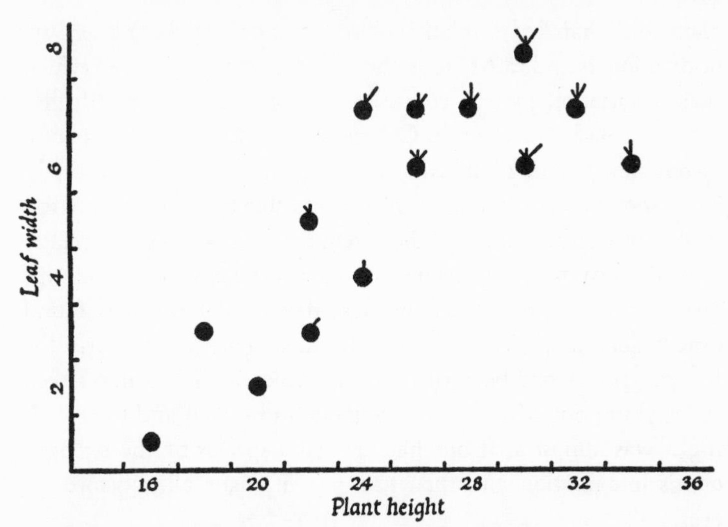

FIGURE 12. The mongrelized collection from the cutover woodland diagrammed by exactly the same scales as were used on *Planta alba* and *Planta rubra* in Figures 9 and 10. Each dot represents one plant, its position on the chart showing its leaf width and its height, while the radiating lines (as in Figures 10 and 11) demonstrate flower color, hairiness of stem, and notching of the petals.

foreign influence can be identified in a key, just as if one had the actual specimen in his hand. It is a laborious process. One studies the variation in a mongrel population item by item, studies it so intensely that it is possible at length to find exact methods of measuring such features as the texture of a leaf, the branching pattern of an inflorescence, the wooliness of a calyx. Then by the use of pictorialized scatter diagrams one gradually works out the degree to which all these various characters tend to more or less hang together. If many of them do (if as in the hypothetical example we found flower color and hairiness of stem and height of plant and shape of leaf all tending to stick together) then hybridization is indicated. It is the only known process by which such a variation pattern can arise. The greatest creative thrill I have ever had was when in Colorado, working with a flora completely new to me, I studied a mongrel population of loco weeds. After some days of intensive work I was able to build up an effective picture in my mind of the foreign species which in a greatly diluted form was responsible for most of the variation in this group of plants, and then the next day I went out and found exactly such a species as the one I had imagined. Of course the foreign species had been there all the time, but I had never seen it. Working out what it must be like item by item and then finding it was almost as if one had created a species by the strength of his imagination and then had seen it materialize before his eyes.

A method like this is obviously one of the new tools with which we may someday hope to understand much about cultivated plants. The method is still a pretty crude one, it requires carefully assembled collections, it needs considerable experience in learning to see and then to measure the kinds of differences which one finds between species, but it can tell us things about avocados and mombins and supposedly wild apples which we cannot hope to determine by other methods.

Though this method is scarcely out of the experimental stage, my students, my collaborators, and I have used it on several cultivated plants—sunflowers, amaranths, avocados, maize, wheat, lilies, narcissi and pigeon peas. It is the study of the avocados that I want to describe in some detail. After I had shown Dr. Williams and Dr. Popenoe of the Escuela Agrícola Panamericana what could be done by this method they were kind enough to make arrangements for me to visit the Rodiles grove of avocados a little to the south of Puebla, Mexico, in the region where several of our most successful American cultivated varieties were originally collected for use in the United States.

It was a fascinating place to work. There was the graceful old white hacienda, now falling into picturesque disrepair. There was the immense volcanic cone of Popocatepetl, close at hand, dominating the western sky line. In the orchard, really a series of orchards, there were old avocados in long straight rows. The Rodiles family have been interested in avocados through several generations. For years they made it a practice to bring home the best ones from the local markets, and to sprout the seeds and plant the seedlings in their orchard. The resulting collection is a botany professor's paradise, literally several thousands of avocados, grown to maturity, no two of them exactly alike. If one wants to study variation in avocados this is it. Some are upright, some spreading. Some have broad leaves as wide as an outspread hand, others little narrow leaves like a pear, silvery white underneath. Some of them reek of anise when one crushes the leaves or opens the fruit; others are quite unscented. Some are late in season, others early. Some have little purple fruits smaller than a sand pear and others have fruits the size and shape of an indoor baseball.

I flew in from Honduras, visited the Comisión del Maíz (going and coming to the Rodiles grove we studied variation in maize, but that is quite another story) which graciously provided me with a pickup truck and a combined chauffeur and scientific assistant. We drove over to Puebla, made arrangements to stay at a neighboring hacienda and got permission from the Rodiles family to study the variation in their remarkable grove. Early the next morning we drove over in the truck. It was a cold dry morning in the winter, the temperature just above freezing; we were thankful for the protected driver's seat as we bumped along over the badly rutted roadway. It was not yet sunrise; the last stars were fading out and ahead of us the perfect cone of Popocatepetl rising Fujiamalike above the flat floor of the valley, was showing a rosy sunrise glow on the snow fields below its summit.

We parked the truck by the old hacienda buildings and walked back and forth the whole length and breadth of the orchard, noticing the variation and discussing how we might measure it. Our breath steamed in the frosty air. We thrashed our arms to bring the feeling back into our fingers; Popocatepetl was now a glowing triangle of dazzling white in the fresh morning sunshine. We stood in a sheltered corner and planned the day's work. The fruits of these avocados we had to leave out of our study almost completely. In the first place they were salable and we had promised not to disturb them. In the second place there was such variation in the fruiting season that only a small proportion of the collection could have been studied at any one time. For our purposes there was enough to study in the leaves alone and we set to on that problem. Though the leaves differed conspicuously from tree to tree, they also differed in size and shape on the same tree and even on the same branch. After a dozen false starts we learned to pick leaves that were mature but which were not from fruiting branches nor from the quick abnormal growth which sprouts out when a tree is pruned or broken. Even with these restrictions there was more variation on the tree than one would have liked. Instead of measuring several leaves and taking the average we accomplished the same result in a much simpler fashion. We selected five leaves from different branches, discarded the two which were most unlike the average of the whole set and then examined the three remaining ones. Usually they were practically identical and we took the one with fewest blemishes for our studies. Occasionally even this method failed to produce consistent results. Then we made three collections of five, reduced each to its average leaf, and then proceeded to study these three leaves and take the most representative. On these carefully selected sample leaves we measured the length and the breadth, the length of the leaf stalk and the angle at which the leaf blade branched out from the stalk. Then we scored the whiteness of the undersurface and the amount of anise fragrance when the leaf was crushed. We then checked our method by having my assistant and myself make samples independently from a group of trees and then studying the precise degree to which my results agreed with his. Gradually the clear Mexican sunshine became warmer; by midmorning it was like Indian summer at home, with fresh cool air and agreeable warmth in the sunshine. We continued, tree after tree, for two full days.

We worked far into the night for a preliminary result on the data from the Rodiles grove. It was worked up into one pictorialized scatter diagram which shows (among other things) that though no two of the trees are alike, much of their variation clusters around two centers. One of these is characterized by trees with small, narrow, long-stalked leaves, gray-white beneath and reeking strongly of anise; the other by trees with large, wide, scentless leaves, green beneath and on very short stalks. The first is rather like the prevailing avocados of Mexico; the second resembles Guatemalan types. Most of the collection, however, is not purely Mexican or Guatemalan but consists of various mixtures and intermediates. From our previous studies of variation in hybrid populations it was evident to us that the collection could have been produced through the crossing of these two very different races of avocado, that the mixture as a whole had much more Mexican blood than Guatemalan, and that few if any of the trees were the original hybrids. To us the data suggested that a long time ago a few Guatemalan varieties had been brought up to this part of Mexico and grown there. They had hybridized with the Mexican avocados and some of these hybrids had then been grown in the area. They crossed among themselves and with the

Mexican types and out of such mixtures had come the superior fruits which the Rodiles family used in setting out their orchard.

Now this one study does not solve the problem of the cultivated avocado but it does point the way to an eventual solution. When groups of wild-growing trees are scored in the same fashion, when man-made crosses between different types are measured in some such way, when all this information has been brought together in one place, then we can sit down with the results and produce a history of the avocado which will be something more than speculation.

One more thing we did before we left the Rodiles grove. From a random lot of trees we selected one leaf from each tree and pressed the whole set carefully. Mounted on a few herbarium sheets, together with the diagram, they give a fairly effective picture of the whole grove. It is a record which can be filed in any herbarium, side-by-side with conventional herbarium specimens. Such herbarium records of variation are one of the features of which I have elsewhere described as "the inclusive herbarium." It is one of the developments in the study of cultivated plants which I shall want to discuss in the concluding pages of this book. Here it will be sufficient merely to point out that by such collections, carefully made and systematically assembled, we may hope eventually to answer such simple but fundamental questions about wild and cultivated plants as those raised in beginning this chapter.

VII
Budgets vs. Scholarship

BY THIS TIME it is clear to me, and I should think it would be evident to the reader, that this book is getting a little out of hand. We have taken up so many side lines, we have considered the study of cultivated plants from such different angles, that about now we need some kind of a summing up to draw the thing together. Suppose, at least for the moment, we stop discussing the contributions of biological techniques to the history of cultivated plants and try to draw up a balance sheet. We find that, considered as a whole, our understanding of cultivated plants in these days is advancing rapidly, if somewhat irregularly and incoherently. Though the taxonomists, as we have seen, have practically deserted the subject which was originally the core of their own special field, this defection has been more than made up by new techniques and new insights which have come in from other fields of biology such as genetics, pollen analysis, and the like. Putting it all together, it is evident that on the biological side we know a great deal more about the history of cultivated plants and weeds than we did fifty or even twenty-five years ago and that significant new information is coming at an accelerating rate.

A rapidly expanding field of study is a fascinating one in which to work. It is not merely that new information comes in. With each new technique, new insights are gained and old facts take on new significance. Before we knew that our bread wheats were polyploids, the fact that *Aegilops squarrosa* was a common weed in ancient Old World grainfields was of little direct interest in the story of wheat. Now that we know that this bristly little weed is actually in all of our bread wheats, we are actively interested in knowing just where and how it grows.

The greatest difficulty is the fundamental one already apparent to a few of us, that the biological side of the problem is only one side. It is quite as much a problem in history, in archaeology, in anthropology, in nutrition, in sociology. It is difficult to find men or groups of men who can and will work effectively in several of these fields at once. Problems which fall straight across departmental and divisional lines run into administrative red tape. A whole series of coherent, fundamental questions are neglected because they do not fall clearly within the domain of any single discipline.

Take for instance a study of the development of the tomato as an important element in the modern diet. It is apparently due to the Italians but the story has yet to be worked out. The biological facts are simple enough and have been brought together in efficient and generally available monographs. The tomato belongs to a South American genus of weedy little plants with small red or orange berries about the size of currants. Out of this complex there somehow arose the cultivated tomato, differing mainly in its increased variability and in its far larger fruit. In pre-Columbian times its great value was recognized in Mexico and for centuries it has been one of the fundamentals in the Mexican national diet.

After the Conquest it was introduced into Europe where it was grown largely as a curiosity. Today it is one of the cornerstones of modern scientific diets. How, where, and when did this change come about? Over most of Europe and in the United States the tomato's virtues were very slow to be recognized. A century ago it was still being grown in small quantities as an ornamental plant for its bright fruits which were frequently said to be poisonous and hence were romantically known as "love apples." In Italy, however, the tomato became so quickly a part of the national diet that it is hard for us to imagine Italian cookery without its inevitable tomato sauces. In Italy the fruit was improved along new lines; Italian peasants either dry tomatoes in the sun for winter use or cook them into a thick paste. New Italian varieties were bred with thicker skins, smoother shapes, and meatier drier flesh and these Italian kinds have been among those used by plant breeders in breeding modern American varieties.

Our own appreciation of the tomato lagged far behind the Italians but in the last three or four decades we have nearly caught up to them. How much of this was due to Italian Americans who brought back to the New World a real appreciation of this life-saving vegetable? In my grandmother's time, tomatoes were grown as a curiosity in the old family flower garden. When I was a small boy they had spread into the vegetable garden. During the season that they were ripe we had them chilled and sliced with vinegar, sugar, salt, and pepper. A few were canned and during the winter months were served as a stewed vegetable about once a month on the average, certainly never more than once a week. A few more were made into chili sauce or catchup, which in our home were typical exotic accessories reserved for company dinners and other special occasions. The catchup bottle had already reached some restaurants but it was not almost universal there, as it is today. By the time I was in my teens tomatoes were eaten in a greater variety of ways. I remember the first time I saw one eaten fresh, like an apple; it was not until after I was married and had a home of my own that tomato juice became a standard item in the breakfast diet. England is twenty-five years behind us, and many of the Continental countries are behind her. Nor is it only in Europe that the special excellencies of the tomato have been slow to be recognized. Around the world in India, the tomato is just now making its way into market places and village gardens. Eventually it should play a great role there. India, like Mexico, is a land of villages with predominantly vegetable diets. Indian climates and day lengths are similar to Mexican and the two cuisines are incredibly alike. In another century the tomato should be as much at home in India as it now is in Mexico.

If we attempt to see this problem of the expanding use of the tomato, to learn how and where and why the Italians led the Western world in finding out that this curious fruit was not poisonous but was tasty and nutritious, we find that there is almost nothing in print on the subject. Certainly as far as modern Italy is concerned, the Italian discovery of the tomato is as important as Garibaldi or Victor Emmanuel. Yet the subject seems never to have been looked into, apparently because it is partly a sociological study, partly historical, partly ethnological, partly nutritional. The problem of the Italian discovery of the merits of the tomato is typical of scores of problems which are either neglected or imperfectly pursued because they fall across departmental lines. If the study of cultivated plants is to advance as it should we must find ways to fuse interdepartmental interests over a broad front. Knowledge is all of one piece but universities (by tradition and for budgetary reasons) are divided into departments.

Similarly, in trying to work out the origin of a cultivated plant we may hope to find part of our clues botanically in studying the origin of the plant, part anthropologically in studying the origin of its uses. One may think of its whole history as a complicated fabric, the warp of which is made up of all the varieties of the crop, the woof, all the various uses to which it has been put. In trying to unravel the history sometimes it will be best to begin with a thread from the warp, sometimes with one from the woof.

Take for example the early history of sweet corn. The botanical facts do not offer us much of a clue. Sweet corns differ from normal varieties of maize by their inability to build sugar up into starch in the kernel. The mature seed, instead of being plumply filled with tightly packed starch grains, is a shrunken, wrinkled irregular mass of dried-down sugar. This is a simple inherited abnormality, a condition reported for several other crop plants as for instance the garden pea. It could apparently turn up in almost any kind of maize once in so many million times. It produces, however, a strange new kernel, one which is more difficult to ripen, more difficult to grind into flour, more difficult to store and to germinate, but one which is definitely sweeter. The North American Indians had found that it made superior green corn and we took it over from them and made it into our most distinctive American table vegetable.

So exclusively have we Americans used sweet corn as green corn (i.e., as a vegetable boiled on the cob before it is fully ripe) that American maize experts just took for granted that only in such ways might sweet corn be used. They established the fact that nowhere in Latin America were sweet varieties used as green corn and then published their conclusions that sweet corn must have originated among the North American Indians. One of these experts sat in a hotel in Guadalajara, Mexico, and wrote out a report that there was no sweet corn in Mexico. From the roof of the hotel in which he sat down to write one can see three Indian villages in which Dr. Isabel Kelly, the ethnologist and archaeologist, collected native varieties of sweet corn. They were not coming into the market in Guadalajara, they were not being used for green corn, but with them as a clue it has been possible to work out the history of sweet corn in the Americas, partly by collecting native varieties of sweet corn and studying them, partly by collecting information as to its native uses. It is a simple and straightforward story if one uses both kinds of evidence.

Sweet corn as a distinctive and appreciated variety apparently originated somewhere in South America. Among the high civilizations of the Andes, some agricultural genius realized that these abnormal, wizzled kernels had more sugar in them. He did not use the new freak for green corn. There are special varieties bred for that purpose in Latin America and of long standing; one of them, for instance, is identical with one of the ceremonial sacred corns of the primitive Huichol Indians of Nayarit, Mexico. In western Mexico there are long, narrow-eared varieties with big blue or red-purple kernels which are widely grown as green corn. Their common name is *maíz de elote,* that is to say, corn which is for eating green on the cob. The ethnologist Lumholtz collected them among the aboriginal Huichol Indians in western Mexico in the early 1890's, when he found them being used as one of the sacred varieties in their religious ceremonies. They make excellent green corn but they do not store sugar instead of starch like our own table varieties. On the background of Latin American corn, which is already fairly sweet, the freak starchless kernels of sweet corn make a product which is unpleasantly gummy when cooked green on the cob. When I asked about it in Mexico I was told "*Ah, señor, se pegen los dientes.*" No, when sweet corn turned up in pre-Columbian highland South America, it was in a civilization which did not have sugar cane. The new product was therefore prized as a source of sugar and used in several special ways. In highland Peru and Bolivia one can still collect the ancient variety of sweet corn which the Incas used in making their high-quality maize beer or *chicha. Chicha* is still a common drink in these regions; nowadays when the fermentation reaches a certain point it is accelerated by adding a little brown sugar. In pre-Columbian times when there was no brown sugar, they added instead a meal made from ground toasted sweet corn to increase the sugar content and give the *chicha* its extra kick. Among a few conservative groups this special sweet corn is still grown and still used as a sugar source in making *chicha.* To eyes accustomed to the long cylindrical maize ears of the American corn belt, this Peruvian sugary corn is a strange-looking variety. The ears are nearly as wide as they are high, as big as an orange, with a thick heavy cob, numerous irregular rows of kernels, tapering to somewhat of a point at the tip and smoothly rounded into a basin at the butt. They may be pale lemon yellow, orange yellow, and various shades of orange red, up to a deep Chinese red.

If one goes northward along the Andes, in primitive communities he finds survivals of various ancient drinks, made out of ground toasted corn, some of them fermented, some unfermented. In highland Guatemala occasional ears of sweet corn have been collected in such communities. Since the brewing of homemade alcoholic beverages is illegal, finding out all the special ways these special varieties are used would take tact, time, and ingenuity. In western Mexico sweet corn is almost a commonplace in little villages and on old haciendas from Jalisco and Nayarit northwards almost to the border. It is called *maíz dulce* and is used in various ways as a source of sugar. The mature kernels are toasted and mixed with peanuts and squash seeds in a kind of primitive crackerjack called *ponteduro.* Or they are made into pinole by being toasted, ground into a fine powder on a metate, flavored with anise or chocolate or cinnamon, and stirred up into a sweetish drink. The pinole powder may also be eaten dry, though I know of nothing drier, and the country people have amusing proverbs testifying that native reaction is the same: "He who has the most spit, eats the most pinole"; "One cannot eat pinole and whistle."

Maíz dulce is widely grown in western Mexico, but it is conspicuously unlike the ordinary maize of that part of the world.

Maize in western Mexico traditionally has a long narrow ear of eight to twelve rows of kernels. It is most frequently white or pale yellow, though as I have said, red-purple and blue-purple varieties are frequently grown for green corn. *Maíz dulce* gives every indication of being directly derived from the ancient South American variety which had spread northward in pre-Columbian times. It still has the same set of colors from old gold to Chinese red. The ears are not narrow like the maize of western Mexico, though they are longer and more tapered than were the South American originals. They still have crowded kernels in irregular rows, they are rounded off smoothly at the butt; they look like the original South American variety with just enough Central American maize mixed into it so that, unlike varieties brought straight from Peru, they can succeed under Mexican conditions. In South America the high row number, the big cobs, the rounded butts, and the varied intensities of Chinese red are common in many other kinds of maize as well. In Mexico they are found only in this one special variety, grown for these special purposes. As one goes northward in Mexico, *maíz dulce* becomes gradually more and more like Mexican maize so that in Sonora and Chihuahua it has ears nearly as long and a cob nearly as narrow as those of other local varieties, though it still retains its distinctive spectrum of unusual kernel colors.

It might be supposed that this distinctive special-purpose sweet corn, spreading up from South America would have been swamped by the common corn of the countries it traveled through. Maize is so naturally cross-pollinated, pollen blows so freely from field to field that this has happened many times in the introduction of new and alien varieties. There are technical reasons why it did not happen to *maíz dulce*. In the language of genetics, sweetness is a recessive character. If a kernel of *maíz dulce* is fertilized with pollen from ordinary field corn, that pollen brings in with it the ability to make starch. That kernel and any others of similar ancestry are as plump and well filled with starch as any ordinary kernel. No observant gardener, primitive or modern, would plant it with his sweet corn. Hence a sweet corn may be grown in the same fields with a starch corn; as long as one takes the precaution of planting only wrinkled kernels no direct crossing can take place.

How then could a little Central American influence have filtered in? By a much more indirect route. If the sweet corn can be fertilized by the field corn, it is equally possible for pollen from the sweet corn to fertilize the field corn, but since the latter carries the dominant starchy condition the crossing will not be apparent. If some of these mongrel seeds (field corn pollinated by sweet corn) were planted as part of the crop of field corn, they would produce hybrids between South and Central American maize and part of their pollen grains would be without the factor for starchiness, though carrying some other Mexican traits. Whenever pollen from one of these mongrel grains fertilizes a kernel of *maíz dulce* it would produce an embryo seed whose inheritance is approximately one quarter Mexican and three quarters of the original type. The kernel looks like a normal kernel of *maíz dulce* but it carries a mixed inheritance. If such a seed is planted, a little ordinary Mexican germ plasm is carried into the sweet variety, and probably produces some plants which are a little more able to survive in Mexico. They are a little more likely to be used for the next crop and as the diluted Mexican influence gets worked into the variety it tends more and more to endow its descendants with increased chances of surviving under Mexican conditions. For these technical reasons an ancient South American variety has been able to move slowly north, century after century, mixing with the ordinary corn of the country enough to adapt itself to the new conditions and yet so protected by its inherent recessivity that in all these years it has not yet lost all of its distinctive South American appearance.

If we now study the sweet corn varieties of the American Indians, we find two more intermediates between our table corn and the ancient *chicha* strengthener of the Incas. Among the Plains Indians are varieties (Nuetta sweet corn, for instance) which are almost like our Golden Bantam except that their kernels are a variety of dilute Chinese reds, some of the same colors as *maíz dulce.* The sweet corn of the Hopi shows even clearer indications of its long journey from South America. This is particularly significant because the Hopi are naturally conservative; due to their isolation they have retained as much of their ancient culture as any Indians in the United States. For their great summer festival they have four sacred varieties of corn, and one of these traditionally is a sweet corn. I have had this sacred sweet corn collected for me in several Hopi pueblos, in quantity from two of them. In color it varies from a pale lemon yellow to yellow overlaid with various shades of red, though none that I have seen quite match the distinctive Chinese reds of *maíz dulce* and the South American sweet corns. The rowing tends to be more irregular than in other Hopi varieties; in some of the collections the number of rows is as great as in *maíz dulce* from Mexico. In many of the ears there is still a strong resemblance to the gently and evenly rounded butt of the original South American variety. On the background of Pueblo maize, the sweet kernels are no longer too gummy to make a good green corn and the Hopi eat their sacred sweet corn on the cob just as we do Golden Bantam and similar kinds. If we list all the ways in which *maíz dulce* is used in western Mexico (*ponteduro,* pinole, etc.) we find that they can be summarized as five special processes. If we do the same with the uses of sweet corn among the Six Nations of New York from whom Americans first obtained their sweet varieties, we can summarize them under seven headings (boiled on the cob, roasted on the cob, grated and scraped, etc.). Now it is a curious fact that all of the seven ways in which the Six Nations use their traditional festive crop are different from the five ways in which sweet corn is employed in Mexico. The Hopi, however, use it in various ways, some of them the same general ways as in western Mexico *plus* some of the ways it is used by the New York State Indians.

When we put all our facts together, the recessivity of sweet kernels, the distinctive kernel colors and ear shapes, the various culinary and ceremonial uses, we emerge with a clear picture. Sweet corn originated in the highlands of South America. It spread northward along the mountains through Central America and western Mexico, after maize was already established there, but in pre-Columbian times. Its original type was gradually changed until finally, apparently somewhere in the Great Plains, in everything but its sweetness it lost all resemblances to the original variety.

With all these kinds of evidence at our command, it is easy to work out a simple coherent history which satisfies botanical, genetical, and anthropological criteria and integrates all the information from these various fields. If we are to make any real progress in understanding the transported floras in which we spend our lives, the work will have to be done by men or by groups of men whose understanding embraces these several disciplines. The relative importance of genetical and ethnological information in working out the history of a crop will vary greatly from one crop or from one variety to another. For sweet corn as we have just seen, the genetic change necessary to produce sweet corn is slight; the changes in the way the new variety is used are tremendous. Therefore, for sweet corn, ethnological data are of primary importance. On the other hand, for Old World versus

New World cottons, the genetic differences are tremendous; a whole new set of chromosomes has been added. The crop itself is virtually unchanged and is used in practically the same ways. Therefore, in cotton the genetical data are of much more primary importance than the ethnological in working out the history of Old World and New World cottons. In the Old World, however, the ancient Indian and African cottons are very similar. It is probable that cotton was used for other purposes before it was taken over finally as a fiber. A careful dredging of primitive communities in Africa and India might bring unique data for interpreting the early Stone Age uses of these ancient plants.

Nor is this true just of the early history of Old World cottons. The importance of fusing anthropological and biological concepts increases tremendously as we work backwards towards the actual origin of a crop plant from its wild or semiwild progenitors. We cannot understand the origin of a cultivated plant until on the one hand we know the plant and the kind of changes which have accumulated between it and its wild progenitors and until on the other hand we understand something of the attitude towards plants and the needs of plants by those who originally domesticated them. There has been far too much armchair speculation by those who knew a little of one of these phases of the problem and there has not been enough careful searching for combined botanical and ethnological evidence which bears directly on particular problems.

There is a close analogy here to studies on the origin of music and musical instruments. The classical armchair theorists told us how man listened to the beautiful sounds in nature, the wind in the trees, the singing of the birds, and then sought to imitate them. Curt Sachs, a realistic exponent of a more modern school, set out with the premise that one cannot understand primitive music without understanding something of the mind and spirit of primitive man. He accordingly spent time in the field studying the most primitive peoples he could find, learning what kind of musical instruments they had, and to what purposes they put them. Among these peoples he found himself in a world of superstition, magic, and taboo. He emerged with the concept that musical instruments originated for magical and religious purposes, like the twanging bowstring, and for signaling, like the drum, and that it was only after there were many musical instruments of various general types already in existence, that very gradually there evolved the idea of combining instruments for the purpose of making beautiful and interesting sounds. Musical instruments are vastly older than instrumental music!

Someone will have to make the same kind of carefully detailed investigations of the attitudes of primitive peoples towards plants before we will be ready to consider seriously the actual origins of our ancient crops. From the little evidence already at hand it is evident that magic had quite as much to do with primitive domestication as with primitive music. Body paints, of magical or religious significance, charms, rattles, magic cure-alls, certainly had as much to do with the origins of food plants as the utilitarian need of food which the armchair experts have stressed in their pronouncements. For all we know now, some of our ornamental plants may have as ancient a history of domestication as any major crop plant. Many primitive peoples plant brilliant flowers around their homes. Inquiries by understanding anthropologists demonstrate that these gaudy plants are not just for ornament, they are for magic: they are scaring away devils. Coxcomb and amaranth are planted in primitive grainfields for the same purpose; who is to know from our present evidence which is the older use, the plant grown for food, or the plant deliberately grown for protection against evil spirits?

A plant which primitive man is known to have held in awe and to have used in his most solemn ceremonies, was not a plant to his way of thinking and not always to ours—common everyday yeast. We still seldom think of it as a domesticated crop, though for nearly a century we have understood that yeast was a living (albeit microscopic) plant. It is one of our oldest domesticates. Modern man uses it for various purposes from brewing and baking to biological engineering and vitamin tablets. Yeast, when grown with a mixture of grain and water, splits up sugar into carbon dioxide and alcohol. In making bread we utilize the bubbling carbon dioxide to raise the dough and let the alcohol escape. In making industrial alcohol we keep the alcohol and do away with the carbon dioxide. In brewing we keep both products; beer gets both its alcohol and its fizz from the activities of yeast.

We do not know when men first began to brew or to bake, but it was a long time ago, and he may well have been a brewer before he was a baker; the histories of brewing and baking are curiously intertwined. Imagine, however, what strange and interesting attitudes a primitive society must have had towards a preparation which could produce alcohol, but only through a set of special incantations. Some of the most primitive yeast sources are various curious mixtures of yeast and fungi and bacteria, which are fairly easily kept alive and may be traded from person to person like a modern yeast cake. Imagine the aura of magic and divinity which must have hedged round these primitive products! They had to be treated in certain ways to be kept potent, they had to be revived by a curious system of adding this and that substance, but when the magic was in them, they transmuted ordinary grain into a beverage which gave one magical power, immunity from discomfort, heavenly visions, as well as headache and nausea sent the next morning by jealous evil spirits. Only a yeast expert and a perceptive anthropologist working together in some of our most primitive existing societies, could get anything like the full story of this fantastic interrelation between yeast, the mind of primitive man, and the origin of civilization.

It is in such fields as this that the barrier of the budget is one of the greatest perils not only to scholarship but to our national culture. At the present time, anthropologists and applied biologists are so far apart in their thinking, that they seldom realize they have any problems in common. When fate throws an agronomist and anthropologist together, in such ways as by giving them adjoining homes in a faculty apartment house, their conversation is usually restricted to general fundamentals — the Brooklyn Dodgers or the personal idiosyncrasies of the Dean of the Faculties. The blame for this deep gulf is about equally distributed. American anthropologists suffer from that inability to distinguish clearly between sophistication and erudition, which once spread cancerously through American universities, so frequently by way of English departments, that I assume Charles Eliot Norton must have been the original affecting source.

Let me give a concrete example of just what this confusion of sophistication and understanding does to a scholar's sense of values. Most of our abler young anthropologists would think that an exhibit at the Museum of Modern Art in New York City was a cultural advantage they ought not to miss if they chanced to be in the vicinity. On the other hand a visit to a corn show in Omaha or Kansas City would never occur to them, and an anthropologist who went out of his way to attend any such exhibition would be looked at with surprise if not with suspicion. Now if one is truly an anthropologist charged with interpreting man and his ways on the North American continent, understanding what corn is in the lives of modern Americans and what it has meant to past civilizations is certainly close to the root of several of his main concerns. This lack of interest in the humble everyday mainsprings of one's own existence is mandarinism. We know mandarin attitudes to have been a sure sign of decay in past civilizations; it probably is such a sign in our own culture. Anthropologists, of all people, should be helping arrest cultural decay instead of speeding the process!

As a further and more specific example of how this dry rot has operated to sterilize anthropological techniques, consider the fate of plant materials in many so-called scientific archaeological excavations. Bushels of prehistoric corn tassels from our own Southwest have been burned by professional archaeologists in their haste to assemble every possible pot and potsherd. They were two thousand years old and more, they were a biological index to vanished civilizations and capable of being analyzed with precision, they were literally just "old corn tassels" to the archaeologists who dug them up and burned them. This burning has now ceased in the Southwest, but the general attitude persists, even in high places. When Junius Bird started back to Peru to investigate the incredible waste heaps of ancient cities in the Viru Valley, one of our greatest American anthropologists told him, "You'll just be wasting your time. I've been all over that site and it isn't worth the trouble." He was speaking not as an anthropologist in the true sense, but as a glorified scientific pothunter. His naturally keen judgment had been warped into confusing artistic excellence and scientific significance. Fortunately Bird was not dissuaded and even though he expected no beautiful pottery, he tunneled into these mounds of ancient litter, preserved for study by a fortunate combination of one of the world's driest climates and salt spray from the adjacent coast. His incredible haul of broken utensils, discarded nets, and thousands of squash rinds, corncobs, and the like, is giving us a definitive understanding of early civilizations in South America.

As for the agronomists and other biological supertechnicians, their minds are quite as sealed to the notion that contact with anthropology might be professionally rewarding. Many of them have a determined yahooism, a resistance to any kind of elegance which is one of the remnants of our frontier tradition.

It would be a healthy thing for scholarship if these barriers between the humanities and the sciences could be broken down more frequently. Scientific work needs background and foreground and perspective; by its nature it must deal intensely with the minutiae of restricted problems. Effective co-operation with the humanities would give it understanding, would lessen this American impulse to do something for the sake of the doing rather than because the end results are desirable. The humanities could benefit quite as much from the interchange. They still operate under the dead hand of medieval scholasticism. Historians, literary scholars, even social scientists, approach many problems as they do, not because that is an effective way, but simply because that is the way which has been followed since medieval times. They have a terrible impulse to begin an investigation as did the scholastics by saying "Now, let us define our terms." The more fruitful scientific attitude is to say, "Here is something peculiar; let us study it. Definitions can wait until we know more about the phenomenon." This is particularly true in biology, where two of our most fundamental units, the species and the gene, are still undefinable in any strict sense, though there has been enough general agreement about them (they can be defined by example if in no other way) so that we work co-operatively towards their eventual definition.

A bringing together of men in different disciplines for the study of cultivated plants and weeds, an active co-operation of historians, anthropologists, and ethnobotanists would have as its immediate aim the advancement of understanding in that particular problem. Its greater ultimate effect would be the catalytic transfer of techniques and attitudes from one field to another.

VIII
Uneconomic Botany

THE GERM OF THIS BOOK, if anything so discursive can be thought of as stemming from a single source, was a course in Economic Botany at the Bussey Institution of Harvard University under Professor Oakes Ames. Since he scarcely mentioned wheat, said nothing whatever about lumber, and gave only a passing nod to such staples as rice and maize, though devoting a whole lecture to amber and spending a full month on arrow poisons, I, in those days, joined the other unappreciative recipients of these pearls in referring to the course derisively as Uneconomic Botany. I have since learned better.

One of the main things I want to discuss in this chapter is the pressing economic importance of some of these matters which I once considered so uneconomic. Before we plunge into such a discussion it might be well to stop by the way and talk a little about Professor Ames. He was one of the few people in this country to take a really intelligent interest in cultivated plants and in his later years he published a small book, *Economic Annuals and Human Cultures,* which is just now beginning to find an appreciative audience. If a scientist is one jump ahead of his fellows in his thinking he is usually their acknowledged leader; if he is two jumps ahead he is thought to be eccentric and rather screwball but sometimes receives belated recognition in his old age. If he is three jumps ahead he is ignored, though posterity may eventually get around to appreciating his evidence as it did with Gregor Mendel. Ames was well ahead of his time in some of his ideas. That and his austere avoidance of anything which might approach proselytizing have kept him from the wide recognition he deserved. Rigidly trained as a true aristocrat, he lived on in a world which was rapidly devaluating its old aristocracies in one way or another. Even among his closest scientific friends there were only a few who realized the extent to which, as a matter of principle, he strictly avoided all quasi-political methods of advancing his own scientific reputation. Not for him the pat on the back here, the nod there, the assistance in getting fellowships, the trading of opportunities for one's graduate students. Not for him the tactful encouragement of younger men likely to advance his ideas. I once chanced to be present when a junior colleague, just back from Washington, was gleefully describing the logrolling which had gone on between the country's scientists in setting up the National Research Council on a permanent basis. Ames listened with growing impatience and finally rose from his chair, tall, thin, and austere. He stood quietly, flicking his folded gloves against his left hand in a gesture of disdain until there was a pause in the conversation. He then looked his colleagues coolly in the face, saying firmly in a cold voice, "If this is Science in America, I will have none of it," and walked out of the laboratory.

Even in those days I was fascinated by his course and defied the tradition that one must not prolong a session by asking the professor any more than the absolute minimum number of questions. I asked so many as to be unmercifully razzed by my fellow students but at the end of the term Professor Ames made a little speech and said there had been such interest in the course this year that he was happy to present all of us with personal copies of Willis's useful little dictionary of plants and their products. (Yet, and this is typical of the man, when I returned to Harvard ten years later to serve under his direction, he remembered neither my name nor my face, nor that I had ever been in one of his classes.)

Ames had the mind of a scholar with the soul of an artist. Anything he ever did was done to perfection down to the last trifling detail, but it was done to perfection by his own aristocratic standards and not according to the conventions of his colleagues or his students. In most years the course attracted a weird mélange of different types of students: a core of hard-working, practical-minded graduate students in plant breeding and economic entomology who seldom bothered to conceal the disdain of young men who have made their own way for someone like Professor Ames who had been born to great wealth; one or two Harvard playboys who needed a few extra advanced credits so badly that they were willing to travel across Boston twice a week to the old Bussey Institute Building, secure in the knowledge that if they just came to the lectures they would get some kind of a passing grade; a few anthropology majors trying desperately hard to be as bored and offhand as was then the fashion for young anthropologists; very occasionally a receptive and appreciative mind like Donald Peattie's.

When I finished the course I thought it the most fascinating one I had ever taken but in my self-sufficient young manhood I was certain that it was also the most useless. Now after three decades of service in botanical gardens as a sort of liaison officer between botany and the general public, I realize that for my purposes at least, it was the most practical.

Though Professor Ames scrupulously avoided even hinting at the fact, he was technically much better equipped than his self-sufficient graduate students to judge just how far his course in Economic Botany dealt with matters which were of truly economic importance. We were just biologists; he was a biologist and a scholar but he was also by necessity a financier. In a period when other Boston families of great wealth settled slowly down towards a barely affluent status, he was shrewdly balancing his investments. He realized far better than we could that modern technology finds new uses for ancient plant products as frequently as it replaces old ones. He told us about balsa wood, the first time most of us had ever heard the name. Five years later it was important in world trade. He went into arrow poisons in great detail; since then a great commercial drug concern has invested a small fortune in bringing an arrow poison, curare, to the practical service of modern medicine, and has done it as a hardheaded practical business venture. Ames told us about fish poisons. We thought it amusing but certainly of no practical significance for us; within a decade such fish poisons as derris root were in mass production in various parts of the tropics to provide one of the most important ingredients in modern insecticides. It was primitive man, Ames told us, who found the strange usefulness of most of our drug plants, all that modern science and technology have done is to be greatly more expert in extracting the drugs, far wiser in understanding and using their effects on the human body. There are five natural sources of caffeine: tea, coffee, the cola plant, cacao, yerba maté and its relatives. Primitive man located all of these five and knew that they reduced fatigue. Biochemical research has not added a single new source.

It seemed to me then that Ames's concern with arrow poisons and ancient crops sprang from the amateurish enthusiasm of a wealthy dilettante. I now realize that his concerns were practical in the strictest sense, though their practicality was concealed by the cultivated urbanity that we youngsters neither appreciated nor understood. "A practical man is a man who practices the errors of his forefathers." It was not a question of who was the more practical; it was a matter of who had the wider view. Ames saw all the plant products of the world as important natural resources. In modern technology an ancient love potion or an obscure fungus are as likely to present us with the great commercial success of day after tomorrow as the plants suggested by narrow-minded practicality.

One of the best examples of the modern technological use of a curious old crop concerns a rare variety of Indian corn known as "waxy" which comes from Asia. It has been used there in various special ways for at least three hundred years, because of its peculiar form of starch. In that part of the world waxy varieties of various cereals are used in preparing gummy desserts such as the delicious fried rice of Chinese restaurants. The carbohydrates in these peculiar grains, instead of being formed molecularly in long carbon chains, are highly branched. Consequently when they are cooked up they form a stickier, gluier substance. Waxy maize was sent to the United States in the early nineteen-hundreds as a curiosity. Geneticists found that in so far as its peculiar waxiness was concerned, it differed from ordinary maize by but a single gene. For forty years it remained as an ethnobotanical and genetical curiosity. It was useful in purely technical experiments in genetics, it became of interest to biochemists when the structure of starches and allied substances was more carefully investigated. At the approach of World War II when a substitute for tapioca was needed for various industrial processes dependent upon gummy carbohydrates, Dr. Merle Jenkins and his associates in the Department of Agriculture transferred the waxy gene to a commercial variety of maize and by the second year of the war, the old-new crop was in commercial production. It was grown and manufactured on a factory scale and maize-tapioca for human food appeared in neat packages in our chain stores as a wartime substitute. Even after the war, with the reappearance of genuine tapioca, waxy maize persisted as a minor element in the national economy. It was a more standardized product than tapioca and therefore more acceptable in certain industrial processes.

With our oldest crop plants, the ones which have traveled with man the longest, multiple uses are probably the rule rather than the exception. We with our modern mass-production agriculture are so used to growing a particular crop for a particular purpose that to us it even seems a little strange to find Italian peasants growing a cereal crop for its grain and also for the straw used in making ladies' hats. The closer we get to primitive agriculture, the more commonly do we find one plant used for various purposes.

It is quite likely that a good many of our crops were not originally used for the purposes which we would now suppose the only reason for growing them. Considered merely as vegetables, the history of squashes and pumpkins is a curious puzzle. All the cultivated species are large and have pleasantly flavored flesh. The wild species are uniformly small and repulsively bitter. How could any such plant have been tolerated even as a crude food, until it developed modifications permitting its use at least as a famine food? Well, in the first place, one has only to live in a Mexican village for a few months to realize that the pumpkin flesh is not the only portion of the plant which makes delicious food. The seeds of many varieties are excellent and on the whole the seeds are almost as important in Mexico as the flesh. So many Americans have appreciated them in the bars of Mexico City that they are now beginning to appear in our specialty shops with other salted nuts. Primitive man, however, had still different needs for pumpkins and squashes. We already know enough about their long and complicated career of domestication to be fairly certain that they started as rattles in ceremonies and dances, and as primitive dishes for eating and for storage. It was probably only long after they had been grown for these purposes and their variability had been increased by the crossing of different kinds, that they began to be used as food, first the seeds and last of all the flesh. Rattle–dish–oilseed–vegetable is probably only the bare outline of these interesting plants which are one of the tripod foundations of the ancient corn-beans-squash civilizations of the Americas.

With some of our crop plants these multiple uses which are a key to their history have lasted into modern technological times. Flax is used both for the oil of its seed (the linseed oil of commerce), and for the fiber of its stem; the commercial crop is mostly planted to varieties deliberately selected either as oil or as fiber plants. Hemp, which gives every indication of having originally joined up with man as a dump-heap camp follower, is a triple-purpose plant. One strain is grown for the oil in its seed, another for its fiber, still another for the powerful drug which in one part of the world is known as hashish, in another as marijuana.

More frequently, however, the earlier uses are all but forgotten. Para rubber has so recently become a world crop that one might suppose its domestication simply the story of a new technical use having been found for a wild plant. Russell Seibert and others who have studied the crop in the field, report a much more complicated story. There is a great variation in the rubber content of the supposedly wild trees; some of the higher-yielding strains trace back to sites which are now part of the jungle but which indicate clearly that they had been village clearings before they were engulfed by the rapidly regenerating tropical brush. From a study of plants from such localities Seibert came to the conclusion that primitive man first domesticated the Para rubber tree for its nuts, if we may use the word *domesticate* to cover such casual use as the deliberate or unconscious introduction of a nut tree into a village. By doing this they brought into their small and transient villages trees which were not just a random selection of the original wild species but just those with superior nuts, or a superior yield of nuts. In doing so they frequently brought in trees which were not native to that immediate area, trees which grew higher on the hills or in even lower sites along the river. As these clearings were deserted the alien trees crossed with those in the immediate vicinity and thus in more than one clearing there eventually developed mongrel swarms of Para rubber trees, which had the heightened variability characteristic of mongrels. It was among them that some of our most potentially valuable breeding material was located when we eventually became much more interested in the milky sap than in the nuts. And so it goes for crop after crop when we make an earnest attempt to look into the history of their uses.

With a little insight it becomes difficult to define the exact boundaries between weeds, cultivated plants, and the truly wild species in the native floras. There are noxious weeds which are actively combated, weeds which are tolerated or ignored, and weeds which are actually encouraged. In western Mexico the prickly poppy and the husk tomato are certainly weeds by any ordinary definition, but their growth is deliberately encouraged to the extent that other weeds will be removed from the maize field but these plants purposely left behind, though they had not deliberately been planted. The fruits from the husk tomatoes are gathered and used in sauces and preserves; in central Mexico there is a larger-fruited cultivated variety of great antiquity which is deliberately planted. The seeds from the prickly poppy are gathered in the late winter and used for their oil and probably also as an opiate, though for obvious reasons it is difficult to get exact information about such practices.

It is not merely its usefulness which makes a crop a crop. More than anything else it is its amazing adaptability. Most plants do not just grow like a radish when you plant them where you please. Only a skilled rock gardener who has tried to bring into cultivation a fair number of the world's wildlings can appreciate the exasperating peccadilloes of the average plant. They want this kind of soil or that, they want just so much shade, or they may prefer shade around the roots but not over the rest of the plant, they must have a certain texture to the soil: not too gritty but just gritty enough. As for germinating their seeds as one does radishes and lettuce, just by sticking them in the soil — that is completely out of the question! Some plants have to get their seeds planted immediately or they die. Others prefer to lie dormant for several seasons and then germinate irregularly a few each year. Others must be kept continuously wet; many of them require freezing and thawing if they are to sprout at all — and so it goes. Most of them are finicky and all of them are finicky in different ways. A plant like the common red radish which can be grown just by putting it in reasonably good soil and keeping some of the weeds away from it, which succeeds almost equally well in the cool fogs of Berkeley or the steaming humidity of a Saint Louis spring, simply does not exist as a part of the truly native flora. This adaptability has to be bred into a crop. Weeds already have a lot of it, and semiweeds have the makings of it. For most of the flora it just isn't there. It is only among some of our garden flowers (a point we shall want to discuss in a later chapter) which have been loved so intensely for their beauty that men were willing to coddle and select, that we have achieved anything like the domestication of wildlings within modern times.

To Professor Ames the indefinite boundary between weeds and cultivated plants, the discovery of our major drug plants by primitive man, the fact that the origin of every major crop is lost in the shadows of prehistory were all clear evidence for a very ancient origin for agriculture. In the five thousand years of recorded history man has not added a single major crop to his list of domesticates. Strawberries, loganberries, and the like have the special advantages of polyploidy and even so some of these crops employ a cultivated fruit as one of the elements from which the polyploid was constructed.

Now, to Ames, who had looked long and carefully into all these matters, there could be little argument about the antiquity of agriculture. If man had so many score major domesticates, if none had been added in five thousand years, then either agriculture must have had incredibly early beginnings, or primitive man as a plant breeder must have been incredibly smarter than modern man. Ames buttressed this conclusion with good hard facts, reviewing in detail the sixty most important cultivated annuals, and published his little book on Economic Annuals and Human Cultures. After reviewing the evidence for European crops, he wrote:

On the evidence of archaeological remains all this was accomplished in the Old World before 3000 B.C., not only for one species but for all the important food annuals. As has already been said, they appear simultaneously as completely developed agricultural species, each one a complex represented by many varieties.

It is reasonable to ask how much has been accomplished in the introduction or creation of new economic species in the last five thousand years, that is, since 3000 B.C. In spite of recent advances in cytogenetical technique made since the beginning of the present century, and notwithstanding the admittedly greater intelligence through which our crops have been made more adaptable to the demands of modern agriculture, it has to be admitted that not a single species comparable to bread wheat, Indian corn, rice or the soybean, or in fact to any of our important food annuals has been added by modern man to the economic flora of the world.

It is also reasonable to ask how long before 3000 B.C. it was that early man began his colossal task of plant amelioration. If a standard for measuring undated time could be created, based on the relative accomplishments of man in antiquity compared with his accomplishments in recent years, it is obvious that five thousand years is too brief a period. It must be multiplied many times, otherwise the period is not long enough for what took place in the evolution of agriculture in the Old World previous to 3000 B.C.

If the wild ancestral forms of our important cultivated annuals were the same as any of the wild species existing today, then it must have taken thousands of years to alter them through cultivation to such an extent that their lineage is no longer traceable. If the lack of wild ancestry occurred for only one or two of these species, some agency other than time might be suspected; but when it is realized that wild ancestry is lacking for all, it becomes very evident that extended time has to be allowed for the disappearance of ancestral forms either through genetical change or through climatic catastrophes, or both.

Turning to the evidence from New World agriculture he continued in much the same vein:

The important American annuals consist of nine genera represented by fifteen species. Taken in systematic sequence they are as follows: quinoa; amaranth; the peanut; Phaseolus with three species, the tepary, the lima and the kidney bean; two species of tobacco; four cucurbits, including the gourd; the sunflower; and maize. Numerically they comprise about one quarter of the outstanding economic annuals of the world, which would seem to be a proper proportion if we take the two American continents as a unit in comparison with the Old World divisions of Asia, Europe, and Africa. Moreover they form a botanically distinct group of species with the possible exception of the gourd and *Amaranthus caudatus*. The American group of "completed species" exhibits all the characteristics indicative of great age on which the botanical time scale of Old World agriculture was based. The multiplicity of races and varieties represented by Zea, Phaseolus, and Cucurbita indicate as intricate a genetical history as that of the Old World genera Triticum, Hordeum, and Oryza.

He compared this conclusion with prevailing anthropological opinion and finally concluded:

Far be it from the botanist to dispute the theories based on sound anthropological evidence of man's origin or arrival in America. No doubt the migrations and discoveries surmised by anthropologists all took place, as did the recorded discoveries of Magellan, De Soto, Hudson, and others. Nevertheless, the hypothesis based on the evidence presented by the enumeration of economic annuals shows that it would have been impossible for wandering tribes, starting from Bering Strait, to travel more than five thousand miles to tropical South America, and discover there the ancestors of a number of useful American plants, and within a period of two or even ten thousand years develop them to the state of perfection they had attained as proved by the prehistoric remains of 1000 B.C. When observed by the first European explorers in 1492, all of these economic species had been diversified and greatly ameliorated, and some of them had been rendered adaptable to every climate from south of the equator to Canada. They had been spread over vast areas of North and South America; they had been rendered dependent on man; they had been so deeply rooted in tribal history that their origin was attributed to the gods. This is too great a task to assign a primitive people in the time allotted. . . .

Biological evidence indicated that man, evolving with his food plants, developed horticulture and agriculture in both hemispheres at a time which may well have reached far back into the Pleistocene.

And there he rested his case.

Unfortunately for Ames, he was a little ahead of his time. The anthropological dogma most fashionable in that decade was that agriculture was a comparatively modern invention, particularly in the New World. It had been pontificated that man had not been in the New World long and that he had had little important cultural contact with the Old World. It was not currently respectable to question either of these postulates. In the decade since Ames's book appeared, new facts, and new techniques making available still other classes of facts, have pushed the estimated dates farther and farther backward. It is now almost universally admitted that man has been in the New World since Pleistocene times. It is too bad Professor Ames might not have lived a little longer to see the tide of scientific opinion turning in his favor.

IX

Dump Heaps and the Origin of Agriculture

WHEN I FIRST WENT to live in San Pedro Tlaquepaque, a small pottery-making town in western Mexico, I was under the mistaken impression that my Mexican neighbors had nothing but dump heaps and a few trees in the yards behind their homes. As I lived there longer and came to know more about the life of the village, I realized that many of these dump heaps were carefully managed gardens and orchards. I saw enough so that several years later in highland Guatemala I played hooky from the cornfields I was supposed to be studying, and spent an afternoon in an Indian village with a Spanish-speaking youth who belonged to one of the two or three Ladino families in that town. Again I came back home with the little more I had learned and thought it over for another two years. By the time of my next visit to Guatemala I was certain that the simple facts concerning these gardens were something worth careful study; and in the few days at my disposal I again visited the village of Santa Lucia on the little height of land between Antigua and Guatemala City and again prevailed upon Señor San Salazar to let one of his sons serve as a guide and interpreter. This time I spent a good part of one day visiting and photographing Indian gardens and finally mapped and measured one of them in detail.

The garden I charted was a small affair about the size of a small city lot in the United States. It was covered with a riotous growth so luxuriant and so apparently planless that any ordinary American or European visitor, accustomed to the puritanical primness of north European gardens, would have supposed (if he even chanced to realize that it was indeed a garden) that it must be a deserted one. Yet when I went through it carefully I could find no plants which were not useful to the owner in one way or another. There were no noxious weeds, the return per man-hour of effort was apparently high, and I came away feeling that as an experienced vegetable gardener (I am one of those strange people who would rather hoe vegetables than play golf or go to the movies) I had gotten more new ideas about growing vegetables than from visiting any other garden anywhere.

The garden, like most of those in Santa Lucia, was rectangular and much longer than broad. On three sides it was surrounded by a low fence made from the local cornstalks, which are so big and so durable that I suspect some of the Guatemalan corns must have been deliberately selected for such purposes. Along the other side was a pruned hedge of chichicaste, a rough-leaved shrub which the Mayas used for various purposes and which has given its name to a well-known Guatemalan town, Chichicastenango, literally, "place of the chichicaste." Growing along the fence and the hedge were several varieties of squash and pumpkins as well as some squash relatives not so well known in the Temperate Zone. There was the perennial chayote whose pale-green, large-seeded fruits are now becoming a common vegetable in our own markets, the related "caiba," much more piquant in flavor, and the black-seeded *Cucurbita ficifolia,* which we know from archaeological evidence to have been one of the most anciently cultivated cucurbits in the New World. The squashes grew more rampantly than they usually do with us and the chayote draped itself over everything, garden walls, trees, mature cornstalks, making the whole garden into a picturesque tangled bower.

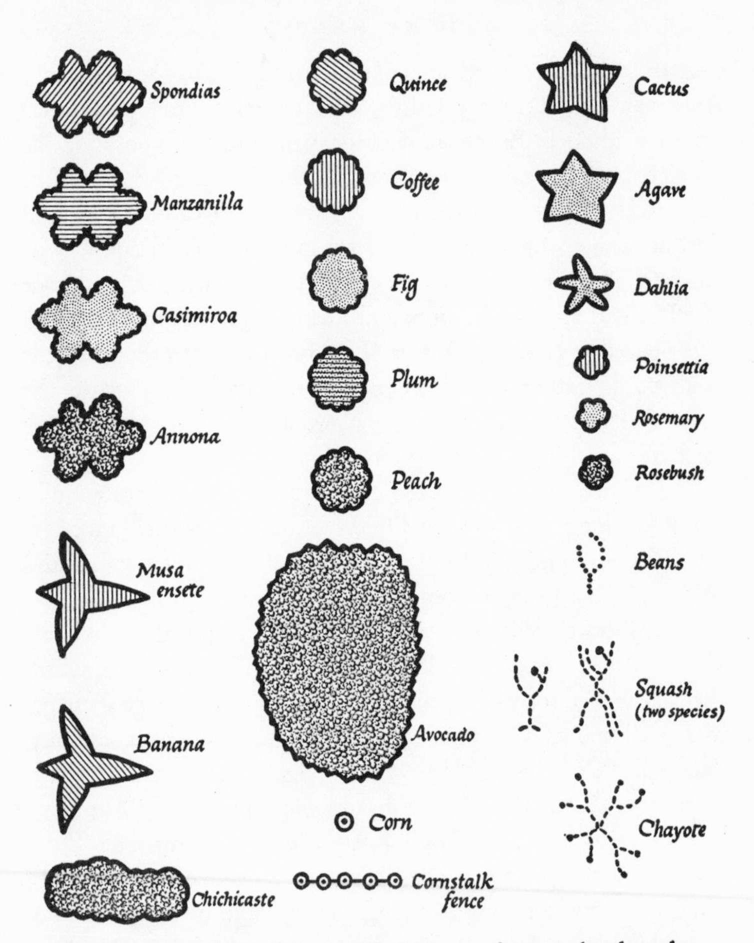

FIGURES 13 and 14. Diagrammatic map of an orchard-garden in the Indian village of Santa Lucia, Guatemala. The glyphs listed above not only identify the plants as shown in the plan on the opposite page, they indicate by their shapes in what general category the plants belong. Circular glyphs indicate fruit trees (such as plum and peach) of European origin; rounded irregular glyphs indicate fruit trees (such as the manzanilla) which are of American origin. Similarly, dotted lines are for

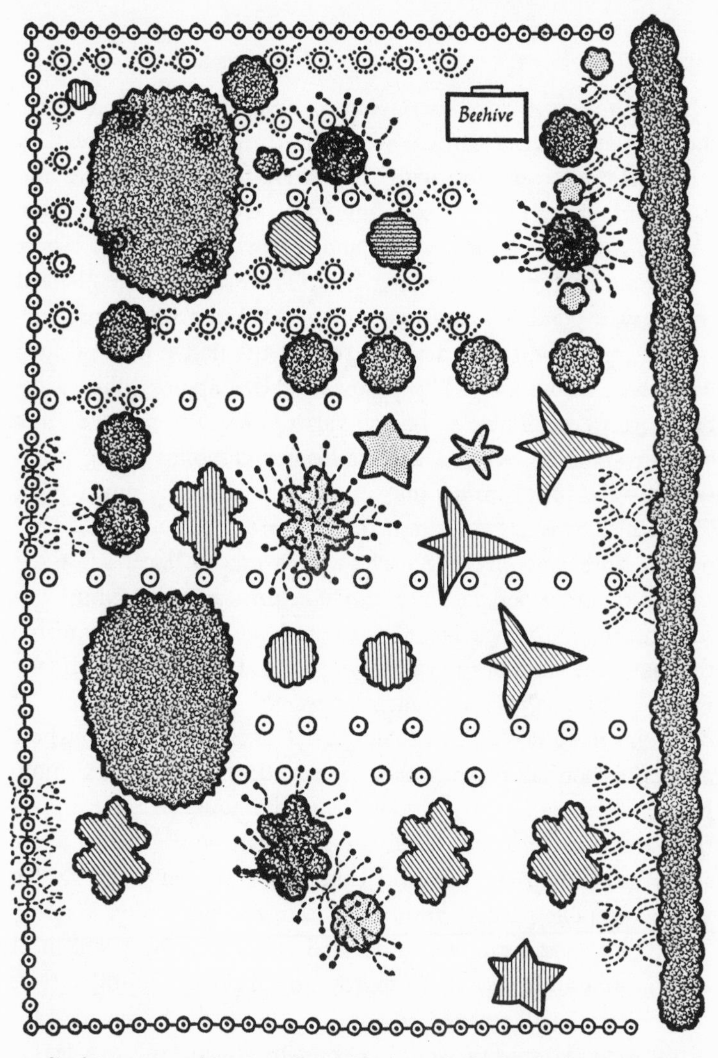

climbing vegetables, small circles for subshrubs, large stars for succulents, and an irregular wedge-shaped figure for plants in the banana family. The long irregular mass at the right-hand side of Figure 14 represents a hedge of "chichicaste," a shrub used by the Mayas.

Though at first sight there seemed little order, as soon as we started mapping the garden, we realized that it was planted in fairly definite crosswise rows. There were fruit trees, native and European in great variety: annonas, cherimoyas, avocados, peaches, quinces, plums, a fig, and a few coffeebushes. There were giant cacti grown for their fruit. There was a large plant of rosemary, a plant of rue, some poinsettias, and a fine semiclimbing tea rose. There was a whole row of the native domesticated hawthorn, whose fruits like yellow, doll-size apples, make a delicious conserve. There were two varieties of corn, one well past bearing and now serving as a trellis for climbing string beans which were just coming into season, the other, a much taller sort, which was tasseling out. There were specimens of a little banana with smooth wide leaves which are the local substitute for wrapping paper, and are also used instead of cornhusks in cooking the native variant of hot tamales. Over it all clambered the luxuriant vines of the various cucurbits. Chayote, when finally mature, has a large nutritious root weighing several pounds. At one point there was a depression the size of a small bathtub where a chayote root had recently been excavated; this served as a dump heap and compost for the waste from the house. At one end of the garden was a small beehive made from boxes and tin cans. In terms of our American and European equivalents the garden was a vegetable garden, an orchard, a medicinal garden, a dump heap, a compost heap, and a beeyard. There was no problem of erosion though it was at the top of a steep slope; the soil surface was practically all covered and apparently would be during most of the year. Humidity would be kept up during the dry season and plants of the same sort were so isolated from one another by intervening vegetation that pests and diseases could not readily spread from plant to plant. The fertility was being conserved; in addition to the waste from the house, mature plants were being buried in between the rows when their usefulness was over.

It is frequently said by Europeans and European Americans that time means nothing to an Indian. This garden seemed to me to be a good example of how the Indian, when we look more than superficially into his activities, is budgeting his time more efficiently than we do. The garden was in continuous production but was taking only a little effort at any one time: a few weeds pulled when one came down to pick the squashes, corn and bean plants dug in between the rows when the last of the climbing beans were picked, and a new crop of something else planted above them a few weeks later.

I was so impressed by the apparent efficiency of the garden that I have since tried out several of its basic principles on my own vegetable plots with considerable success. Instead of putting my sweet potatoes all neatly in one little bed down at the far end of the garden I plant them one row at a time in different places. They now grow out vigorously across the garden by late summer; they keep the ground moist during the dry days of August; and they help keep out weeds. I also plant a few cornfield beans in among the corn plants after they are pretty well up and have a good extra crop of string beans after the sweet-corn season is over. From these experiences I suspect that if one were to make a careful time study of such an Indian garden, one would find it more productive than ours in terms of pounds of vegetables and fruit per man-hour per square foot of ground. Far from saying that time means nothing to an Indian, I would suggest that it means so much more to him that he does not wish to waste it in profitless effort as we do.

This is the most primitive type of garden which I have so far been able to study personally; from the literature I gather that it is characteristic of wide areas in the New World, Asia, and Africa.

Under more strictly tropical conditions (because of its altitude, Santa Lucia for all practical purposes is almost in the Temperate Zone) such garden-orchards blend even more closely into the native vegetation. In Malaysia, on more than one occasion deliberately planted native orchards have been mistaken for part of the natural woodland by European and American plant collectors. Varieties of citrus fruits said to be wild were actually collected from native gardens by Occidental botanists so imperceptive that they could not distinguish between man-made orchards and more-or-less natural vegetation. Perhaps a goodly number of the primitive varieties of various cultivated plants which have been collected at one time or another in the tropics by northern botanists and which are said to be the wild ancestors of such-and-such a crop are in the same category. Here, as in many other basic problems of tropical agriculture, we are up against a serious difficulty. Agriculture, as an ancient art, began in the tropics and has various special complexities there by reason of its long persistence in those areas. Agriculture as a modern science developed in the Temperate Zone. Most of our scientific understanding of agriculture comes from our experiences during the last few centuries with the relatively simple agricultural problems of northern Europe and North America. When the average scientific agriculturist goes to the tropics he has much more to unlearn than to teach, but he frequently seems to be unaware of that fact.

These hit-or-miss tropical gardens are of particular significance because they fit in perfectly with a theory of Carl Sauer's that agriculture may have originated among a sedentary fisherfolk. The first definite beginnings of agriculture took place very early, about the time man domesticated the dog, perhaps in those early Neolithic times which many authorities designate as the Mesolithic. As a geographer, Sauer thinks of the Mesolithic as a time in which rapidly melting ice sheets were raising the sea level all over the world, flooding continental shelves which had been bared when a good portion of the world's precipitation was locked up in polar icecaps. It was a time when coastal valleys were becoming estuaries, when there was a rapid building of new deltas, an increase in length and complexities of shore lines and the formation of river swamps and oxbow lakes. In such an environment people living primarily by fishing would have been able to supply themselves with food without shifting from place to place. The archaeological records show that shell middens accumulated of such a size as to indicate centuries of occupation. Excavations indicate that this was an era in which there was an elaboration of fishing gear, boats, harpoons, and fishhooks. The fishhooks are important for our story; they imply fishing lines, and fishing lines indicate the use of plants for cordage.

From the world-wide distribution of fish poisons Sauer infers that this lazy-man's way of stunning your prey with a plant poison is a very ancient trait. He thinks it likely that the first fish poisons may have followed naturally from the making of fish-lines. What could be more natural than that in the process of bruising vines and bark with blunt stone instruments and soaking the fibers to rot away the soft parts, man should accidentally have discovered that some of these plants can stupefy fish and make them easier to catch? As a further argument he points out that some primitive fiber plants are known to be still in use as fish poisons. To him it seems likely that such a folk, blessed with abundant food, becoming increasingly skilled in navigation and hence in transport, already using plants for fiber and for poison, might gradually shift from plant gathering, to unintentional domestication, to the purposeful growing of plants. He points out that in one of the most ancient centers for man, southeastern Asia, there was extensive drowning of ancient coast lines and the production of long and complicated new ones. In this area he finds evidence for an early center of gathering and preparing plant products with blunt stone instruments, the grubbing up of tuberous plants from river swamps, the manufacture of nets and cordage, the elaboration of fish poisons, the manufacture of cloth and of bark, and the building of bark houses. He brings out the interesting fact that the making of sago by shredding and soaking and pounding the stems of certain palms and cycads is mostly limited to this area. "The practice of shredding, pounding, washing, and decanting runs through plant uses throughout southeast Asia and seems to tie fiber-making and toughening, poison preparation for fishing, hunting, and medicine, and food preparation, including the coagulation of sago, into one culture complex."

To me this theory is worth careful consideration because such people would have also created refuse heaps and I am even more intrigued than Sauer by the notion that such refuse heaps may have played a key role in the origin of cultivated plants. These ancient dump heaps are of extraordinary interest in connection with the origin of cultivated plants, because they are open habitats.

What do we mean by an open habitat? Well, let me begin my explanation by telling you the story of Mrs. Swune and the hollyhocks. Mrs. Swune (that was not her real name, but she was a very real person) loved hollyhocks, and they did well in her garden with very little care. Seedlings came up all through the perennial border and even among the shrubbery, and all during the early summer her place was ablaze with hollyhocks. Naturally, she became very hollyhock-conscious, and she wrote some poems about them and had her friends in for a sedate tea among the hollyhocks, and brought out the poems and read them to her guests. I was invited to one of the teas, and it really wasn't as bad as it sounds: the poems were short and there were plenty of excellent teacakes, and the other guests were swell people and we kind of banded together. At the conclusion of the tea she presented each of us with a large packet of hollyhock seed and wanted to know if we would please scatter them through the Ozarks whenever we drove down that way. Spring bloom in the Ozarks she found satisfactory, but to her the summer roadsides were dull, and she thought if we would just toss the seeds out the windows as we motored along they would just naturally come up all through the Ozarks, and Highway 61 would become a Hollyhock Memory Trail.

Well, I knew the hollyhocks wouldn't come up (even though they do grow along roadsides in the Eastern states), but it is fruitless to argue with people like Mrs. Swune. She was so persistent that eventually I even tossed my seeds out the car window according to her directions, and so, I suspect, did some of the other guests, but the Ozark highways are just as bare of volunteer hollyhocks as they ever were. What Mrs. Swune did not understand is that most plants are very choosy about where they will and will not grow, and that some places, like gardens and dump heaps, are relatively open habitats, receptive to a good many kinds of plants, while other places, such as meadows and mountaintops are relatively closed habitats in which aliens will have trouble getting a footing.

A useful concept in discussing such problems is that every organism has a kind of niche which it has been evolved to fit into (an ecological niche is the precise scientific way of phrasing it); and that many of the things which happen in the domestication of plants and the origin of weeds can best be understood in terms of this concept. It really isn't so complicated when you start thinking it over. Even Mrs. Swune would have realized that you can't plant orange seeds in Greenland and expect to raise orange orchards. Any person of normal intelligence knows that all plants and animals have different likes and dislikes; some are suited to one place and some to another. You don't try to grow water lilies in a desert or plant cacti in a redwood grove. What such people as Mrs. Swune do *not* realize is not only that oranges will not do in Greenland, but that the average plant is most awfully finicky about just where and when it will grow, under exactly what conditions it will germinate, under exactly what circumstances it will persist to maturity after it has once come up as a seedling. So choosy are most plants about these matters, so individual are most of their likes and dislikes, that we understand it scientifically in only the crudest sort of way. Precise limits of temperature and moisture have been worked out for some plants but the vastly more intricate business of which plants they will and will not tolerate as neighbors and under what conditions, has never been looked into except in a preliminary way for a few species.

If you don't mind, let us again consider the spiderworts. You will remember that *Tradescantia ohiensis* and *T. pilosa* seldom produced hybrids in nature, though in the experimental plot they came up like weeds. We interpreted this as due to the garden being an open habitat. The complex flora of the woodland was made up of species which had evolved together. Natural selection had made them fit into each other's company like the pieces of an intricate jigsaw puzzle. If the plants in this association went to seed, the seedling was likely to find a niche to which it was suited and its chances of survival were good. If a hybrid seed, or an alien seed from some other flora were planted there, even though able to germinate, it did not fit into the strict interlocked economy of that vegetation, or the chances of its doing so were miniscule. The habitat was closed. Start making a refuse heap in the neighborhood of these two species, however, and one might find even more hybrids in it than in a garden. Here is a strange new kind of habitat. Many of the plants in the native flora do not fit into it, some aliens will, and some hybrids. Plants which can grow in such places will have less interference from other plants. Kitchen middens would be likely places in which fruit pits, seed heads, and the like, brought to the village from some distance, might germinate and survive. However, before we come to grips with the dump-heap theory, let us spend a little longer with the concept of the ecological niche.

I learned a great deal about such matters from seeing a fire line plowed across upland meadows in our Arboretum. In the lowlands of this area, meadows and fields are dotted every spring with the green-yellow blossoms of winter cress, a kind of wild mustard. With us, the winter cress is common in the flood plain, but until we plowed the fire lines it had never appeared on our hillsides. Yet for several years after the plowing was done, there were so many winter-cress plants along the fire line that they made a streak of yellow in the landscape. Since then the grass has grown back in and the cress is again restricted to the lowlands. From this I learned that cress grew down in the lowlands not just because it wanted more water, but because the water in the lowlands held in check some of the plants, such as bluegrass, which competed with the cress. Standing water made enough bare spots in the flood-plain meadows so that cress got started there; on the uplands, it required plowing to give cress a start and repeated plowing if it was to persist.

Nor is winter cress the only lowland plant which one finds in the uplands, following a plowing. If we watch deserted fields in the northern edge of the Ozarks, most of the trees which come up spontaneously are flood-plain trees, rather than upland trees. Cut down a woodland of oak and maple, or of oak and hickory. Clear the land and plow, grow wheat, or soybeans, or potatoes for a while, and then leave the field fallow. Even though there are oaks and maples and hickories all around the edge of the plot, they are not the first trees to come up there. Far from it. It is elms and sycamores and honey locusts, all of them trees which belong in the flood plain.

Why do flood-plain trees come up in deserted upland fields? Drive around Missouri in the wintertime and look at the sycamores and you will find your answer. The sycamores are easy to spot in the landscape. Their bark peels off in big flakes and in the wintertime they are nearly as shining white as a paper birch. In Missouri they are one of our commonest trees and one sees them lining the banks of rivers, forming little parks of all one kind of tree on our big gravel bars, or growing in the actual beds of small creeks. Ordinarily they are seen only along these watercourses and you might easily suppose that only there do they find things to their liking. On a day's drive, with sharp eyes, one can find a fair number of exceptions. These are the key to the story. Quite frequently one sees a few isolated sycamores, far from any watercourse, in a feed lot or near a barn. Some few of these may intentionally have been planted there, but surely not all of them. In the abandoned mine dumps and borrow pits of the lead belt and the tiff district one sees much more evidence of this sort. These picturesque small-scale mining operations have been going on for over a century. On brushy hillsides, irregularly dug-over fifty to a hundred years ago, sycamores are nearly as common as they are along the rivers and creeks. The first need of a sycamore seedling is evidently not a high water table; it is open soil. On the uplands under natural conditions, areas of exposed soil are very rare; it is only when a big tree blows over and brings up a mound of soil with its roots that one finds them at all. In the lowlands the river is continually plowing its banks and dumping mud, sand, and gravel in new places. Upland plants have not been evolved to fit into habitats where the topsoil is being churned and rechurned; many flood-plain plants like nothing better. So it was that flood-plain trees came up in the abandoned upland fields.

Not until they had grown to some size and made a leaf-cover over the raw soil would the sugar maples and white oak which belonged there spread back in.

Now patches of open soil, like dump heaps, are a part of our story, for these are two of the commonest scars man leaves on the landscape. When he began to spread out of his original corner and into lands previously without human inhabitants, the open habitats which he tended to create, the strange new niches where something different might get a foothold were dump heaps and patches of open, more-or-less eroded soil. In both of these habitats his only natural partner was the big rivers. They too make dump heaps of a sort; they too plow up the mantle of vegetation and leave raw scars in it. Rivers are weed breeders; so is man, and many of the plants which follow us about have the look of belonging originally on gravel bars or mudbanks.

If we now reconsider the kitchen middens of our sedentary fisherfolk, it seems that they would be a natural place where some of the aggressive plants from the riverbanks might find a home, where seeds and fruits brought back from up the hill or down the river might sometimes sprout and to which even more rarely would be brought seeds from across the lake or from another island. Species which had never intermingled might do so there, and the open habitat of the dump heap would be a more likely niche in which strange new mongrels could survive than any which had been there before man came along. Century after century these dump heaps should have bred a strange new weed flora and when man first took to growing plants, these dump-heap mongrels would be among the most likely candidates.

If we look over our cultivated plants with the dump-heap theory in mind, we find that a goodly number of our oldest crops look as if they might well have come from some such place. Hemp is difficult to keep off of modern dump heaps once it is established in a neighborhood; squashes, pumpkins, beans all have the look of such an origin. Among the world's oldest but least-known crops are the grain amaranths. These big coarse pigweeds, or "redroot" as they are sometimes called, have little seeds about the size of the head of a common pin. They are grown as a grain crop by primitive highland peoples both in the Old World and the New. The red-leaved variants are used for food color, or planted in other crops to scare away devils; the young leaves are used for greens; the seeds are made into a gruel or are popped and used in foods and various religious rites. Amaranths are a dump-heap plant par excellence, and are common in barnyards, middens, and refuse dumps throughout the world. The ancient Aztecs in a sort of pagan communion ceremony mixed the popped seeds with human blood, molding the mess into the shape of a god which was sacrificed on the altars and then passed around to be eaten. Of these grain and weed amaranths we can write with some assurance, for Jonathan Sauer has begun their careful study; as the result of his work we have a fairly precise notion of what species of grain amaranth there are, where they are grown, and how they are used. For an understanding of many other dump-heap plants we shall have to wait until other scholars take an equally keen interest in the other humble plants which have traveled with us so long and so far. The history of weeds is the history of man, but we do not yet have the facts that will let us sit down and write very much of it.

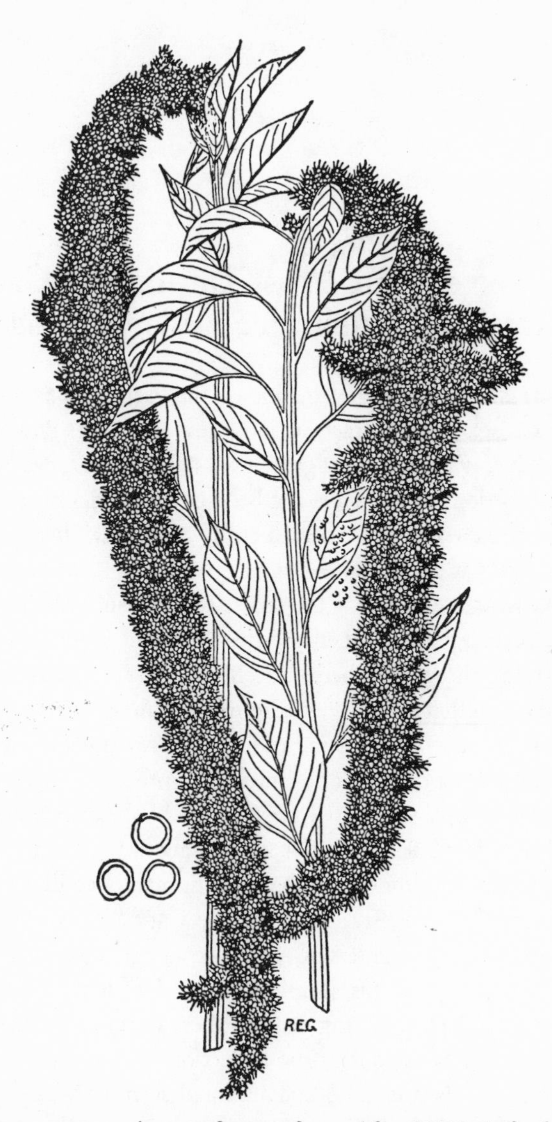

FIGURE 15. *Amaranthus caudatus*. After W. E. Safford

X

A Roster of Our Most Important Crop Plants and Their Probable Origins

Science, as I have suggested in one of the earlier chapters, is a very poor housekeeper. In fact, in some respects she is downright disorderly. Her vast and ever-growing establishment has rooms in its older portions which have not been put to rights for decades and in which there has never been a really systematic dusting. For a really important suite, those rooms given over to the study of cultivated plants are in about the worst state of any. Modern discoveries in cytology and genetics have produced wonderful new equipment for these chambers, but Madame Science is such a frowsy old dame that she has done no more than to open the door, set in the shiny new equipment, and then hurry quickly away.

All of this poses a difficult and immediate problem. By the very nature of the subject we must have for this book some kind of a general review of what is or is not known about the origin of our cultivated plants. If I knew less about the subject, I could authoritatively quote the great Swiss botanist de Candolle, who last set this suite of rooms to rights some sixty years ago and let it go at that. Unfortunately for this simple solution I have spent a good deal of time recently in poking around these dusty chambers, trying to make some sense out of the collections there, and I realize what a mess the place is in. A conventional scientist would probably do one of two things. Either he would keep his sleeves rolled up and put in two or three decades of good hard work, or he would close the door discreetly and tiptoe softly away. In neither case would he ordinarily mention the scandalous condition of the suite to any of the general public. Well, I for one think it is high time this convention be challenged, and the general public informed about the deficiencies of science as well as its glories. Science has been developing a tendency to get pretty high-handed. It will do no harm for once to discuss its shortcomings along with its accomplishments.

The following review of around one hundred of the world's most important and most significant crops is therefore the kind of thing which ordinarily is kept strictly within the scientific family. Each crop deserves careful monographic treatment; for most of them scarcely even a beginning has been made. For a few of the most important crops such as cotton and wheat, a series of monographs have shown that the problem of the origin of cultivated plants is immensely bigger than it seemed to de Candolle and the botanists of his day. It will take a series of monographers merely to outline the job he was once thought to have finished fairly acceptably.

Writing up the review presents us with an editorial problem, one which has been facing us all along but which we have managed to sidestep fairly effectively. This book has been written for the general public; for the intelligent public to be sure, but with no thought of restricting its use to those technically trained in biology. The public will want the general picture and will have little interest in where the facts came from; any scholar who chances to read this work may have little or no interest in the general picture, but he will want to know down to the last finicky detail where the facts came from! As a concession to this ruling passion (after all, I suffer from it myself) I have appended to each item the briefest of references. While too cryptic for most students they will be adequate for most scholars. Much of the information in the review comes from personal interviews with plant breeders and ethnobotanists, from trips to native markets and to village fields. The notation *EA* signifies that in part at least the discussion of that particular item goes back to this private stock of information. Where my own observations have been published I have cited my own papers like those of any other author.

In assembling the roster, crops have been alphabetically indexed under their common English names in as broad categories as possible. Crops which are used in more than one way (as for instance hemp, which is an oil plant, a fiber, and a drug) have been indexed under whichever seemed the most convenient. A complete cross referencing can be obtained by using the general index at the back of the book. It includes both common and scientific names.

Beverages, Condiments, and Drugs

Cacao: *Theobroma cacao*

The source of our chocolate and cocoa was extensively cultivated in the New World by the time of Columbus. There are several species in the genus and cacao supposedly originated in Central America through hybridization between two or more of them. *EA*

Castor bean: *Ricinus communis*

This weedy plant runs wild in clearings, roadsides, and dump heaps throughout the tropics and subtropics. It probably originated as a dump-heap camp follower which was gradually developed as an oil plant, a drug, and an ornamental. *EA*

COFFEE: various species of *Coffea*

There are forty species of Coffea native to Africa and India, more than one of which was used from very early times to reduce fatigue, usually by making a tea from the leaves. The most important species, *Coffea arabica,* has been used in Ethiopia since very early times, though the use of the roasted seed for beverage purposes is comparatively modern. The relationships of these various species and the extent to which they have been consciously and unconsciously modified by man remain to be worked out. *Cheney, 1925.*

OPIUM POPPY: *Papaver somniferum*

Widely cultivated since the Stone Age. Probably derived from the closely related *Papaver setigerum* of the Mediterranean region. The drug is obtained by pricking the ripening seed heads and allowing the milky sap to dry. In commenting on the early domestication of this and some other drug plants, Professor Ames remarked, "When the elaborate methods of preparation of some of the plants used to break down the monotony of life are studied, it becomes quite evident that primitive man must have possessed something other than chance to reveal to him the properties of food and drug plants, to discover the secrets of fermentation, the effect and localization of alkaloids and toxic resins, and the art of roasting or burning a product to gain from it desired narcotization or pleasing aromas (coffee). To fermentation and fire, civilizations owe a tremendous debt." *Ames, 1939; EA*

PEPPER: *Piper nigrum*

Both the white and black peppers of Occidental use come from the fruits of this species. It is one of a number of species of Piper which have been used as condiments in the Orient since very early times. Reports that *Piper nigrum* occurs wild in the foot-hills of the Himalayas are the result of confusion with other peppers. It does grow without cultivation in the hills of Assam, but whether native or run wild is not clear from the evidence at hand. *Burkill, 1935*

TEA: *Camellia sinensis*

A beverage of great antiquity in the Orient, where it was apparently preceded by the direct chewing of the leaf to relieve fatigue. Modern studies of the classification of wild and cultivated teas have progressed far enough to indicate that nothing authoritative can be written about its ultimate origin until much more critical work has been done and living collections for study have been assembled. *Burkill, 1935*

TOBACCO: *Nicotiana tabacum*

There are two different cultivated species of which this is by far the more important. It is known to be a true-breeding polyploid hybrid of two semiweeds from South America. It undoubtedly arose there, probably in cultivation, in early pre-Columbian times. *Schiemann, 1932; EA*

INDIAN TOBACCO: *Nicotiana rustica*

Hardier than *Nicotiana tabacum* and the species most commonly cultivated by the Indians of North America. Like the common tobacco it is a true-breeding tetraploid hybrid between two South American species and must have originated, presumably in cultivation, in South America. *Schiemann, 1932; EA*

TURMERIC: *Curcuma domestica*

This relative of the ginger is used in the United States only for pickling spice and to flavor and color curries, but it may be one of the world's oldest cultivated plants. It is widely and variously used in India and the Malay Peninsula, where there are various other species of the genus, several of them also being used as spices and for the coloring matter in their tuberous roots. It was once widely used as a body paint, a practice which still survives in Indian villages, where it is used as the source of the bright vermilion caste marks worn on the face by orthodox Hindus. Its ceremonial use as a skin conditioner and a female symbol is widespread in the same area. It is not known in the wild state and the cultivated form has never been known to set seed. All of these facts put together suggest that turmeric's long association with man may date back to the very beginnings of agriculture. *EA*

Ornamentals

DAHLIA: *Dahlia variabilis*

Cultivated in Mexico since very early times. The cytological evidence shows that it is a true-breeding polyploid hybrid between two quite different kinds of dahlias. This hybridization most probably took place in cultivation after at least one of the parental sorts had itself been domesticated. *Lawrence, 1930*

MARIGOLD: various species of *Tagetes*

These common ornamental plants are probably among our oldest domesticates. In one form or another they were the ancient sacred flower of the Aztecs and the ancient sacred flower of the Hindus. They form a polyploid complex and have been widely cultivated in various parts of the world. Nothing can be said about their probable origin or their history until their relationships to each other have been carefully worked out. *EA*

Root Crops

CASSAVA: *Manihot esculenta*

One of the greatest of tropical food plants, variously known as cassava, manihot, tapioca, and yuca. Originated in South America in early pre-Columbian times. Both poisonous and non-poisonous races are known. A careful study of the classification of the races and the various uses to which the roots and other parts of the plant are put would tell us a great deal about the early history of agriculture. *Burkill, 1935; EA*

POTATOES: *Solanum tuberosum*

In the last twenty-five years special collecting expeditions sent to the New World by the Russian and by the English governments have completely revised our understanding of the origins of the potato. There are several scores of species and varieties of the potato from Mexico along the mountains to Chile. There are weed, wild, and cultivated potatoes, the whole group forming a gigantic complex with hybridization and polyploidy to increase its complications. In part, at least, this complex is prehuman. In part it is certainly the result of man's conscious and unconscious effects upon evolution in Solanum. *Hawkes 1944; EA*

SWEET POTATOES: *Ipomoea batatas*

An ancient American crop which originated in some unknown way from the widespread American genus *Ipomoea.* Now widely grown in the tropics, its spread was the result of two waves of introduction: (1) an early one by the Polynesians which carried it as far as New Zealand, which it is known to have reached in pre-Columbian times; (2) post-Columbian introduction by the Spanish into the Philippines, where the Mexican name is still current. The ultimate extent of these two waves into southeastern Asia has not been critically established. *Burkill, 1935; EA*

TARO (elephant's-ear): *Colocasia*

The innumerable cultivated forms and the various species of this genus are wild, run wild, and cultivated over wide areas in southeastern Asia and Polynesia. Spier concluded, on summarizing the existing evidence, that the cultivated varieties probably originated in Assam or upper Burma. *Spier, 1951*

Grains

BARLEY: *Hordeum vulgare* (frequently treated as two or more species)

Probably the oldest of our major cereals. Widespread in Neolithic times. There are two centers of diversity, one in Abyssinia, the other in Nepal and Tibet. A weed barley, *Hordeum spontaneum,* once thought to have been the ancestor of cultivated barley, is now known to be distributed from Morocco to Abyssinia, Asia Minor, Persia, and Turkestan. Its relation to the various cultivated barleys has yet to be worked out. *Schiemann, 1932; EA*

BUCKWHEAT: *Fagopyrum esculentum*

Probably originated in central Asia. It reached Europe overland in the late Middle Ages. *Ames, 1939*

GRAIN AMARANTHS: *Amaranthus caudatus* and several other species

Jonathan Sauer's ethnobotanical monograph on the grain amaranths has brought order where previously there had been much confusion. The seeds of these giant pigweeds have been an important crop since very ancient times and the cultivated species are all distinct from the wild species. They are used as grain, as a leafy vegetable, for food coloring, for ornament, and for magical and religious purposes. Sauer demonstrated that those used for grain in the mountains of the Orient are identical with two of the American cultivated species. *Sauer, 1950*

JOB'S-TEARS: various species of *Coix*

The glassy-shelled little kernels which we know only as a curiosity or as material for rosaries and necklaces are part of a polyploid complex which includes an important tropical cereal. The edible forms lack the stony covering. Tropical varieties grow luxuriantly and are larger than maize plants. The whole complex is native to southeastern Asia though little has been done as yet to separate the cultivated varieties from the weeds which gave rise to them and the weeds which they produced. *Watt, 1885; EA*

MAIZE: *Zea mays*

An abundant archaeological record in the New World dating back to before 1000 B.C. in both North and South America is incontrovertible evidence that maize was native to pre-Columbian America. However, the presence of distinctive strains and distinctive uses of maize among aborigines in southeastern Asia raises for some of us the question of whether it might have crossed the Pacific in very early pre-Columbian times. The only close relative of maize is teosinte (*Euchlaena mexicana*), a weed in fields and abandoned areas in Mexico and Central America. It is almost certainly derived from crosses between maize and Tripsacum, a native American grass, and therefore originated from maize rather than being its "wild" ancestor as has been very generally supposed. *Anderson, 1947*

OATS: *Avena sativa*

Even more than in the case of wheat, new cytological and genetical studies have thrown previous theories of the origin of oats into a cocked hat. Our cultivated oats are diploids; most of the ubiquitous "wild oats" of California, the Mediterranean, and many other parts of the world are a series of related polyploids. This means that the crop *could* have helped give rise directly to the weeds, but not the weeds to the crop. Our cultivated oats apparently had a long history as a weed in fields of emmer, the primitive tetraploid wheat of the Neolithic and later times. Gradually the weed was tolerated and eventually became a crop in its own right. But where, when, and how did the diploid weed which spawned our diploid oats first come into being? *Schiemann, 1932*

PEARL MILLET: *Pennisetum glaucum*

One of the "minor millets," which has been grown in India since ancient times, as well as in Africa, where it is supposed to have originated. *Burkill 1935; EA*

QUINOA: *Chenopodium quinoa*

One of the great crops of the New World in pre-Columbian times and still important at high altitudes in South America. It is a close relative of our lamb's-quarters or smooth pigweed, and various other species have been domesticated or semidomesticated. The tiny seeds are borne in great abundance and are used for porridge or ground into flour. At high altitudes quinoa probably produces more bushels of grain per acre than any other crop. We shall know virtually nothing about the relationships and origin of this ancient crop until someone straightens out the confused and neglected classification of our New World species. *Ames, 1939; EA*

RAGGEE MILLET: *Eleusine corocana*

One of the ancient "minor millets" of India. Extensively grown on poorer sites in the Deccan. Closely related to the common weed, goose grass, *Eleusine indica.* Since it is a tetraploid it might be a true-breeding hybrid of the latter and some other species, for no wild form is known. It would seem reasonable to suppose that Raggee is very ancient as a crop, since supposedly it would never have been domesticated after superior grains were available. *Porteres, 1951; EA*

RICE: *Oryza sativa*

Rice is certainly of great antiquity in southeastern Asia. However, the presence in that same area of such apparently indigenous primitive crops as the minor millets and Job's-tears would lead one to suspect that even over much of that area it was preceded by other cereals. There are widespread weed rices whose relationship to cultivated rice needs to be carefully determined. Some of them are tetraploid, indicating that they might be derivatives of rice rather than ancestors, since rice is a diploid. It has recently been found that the Japanese and Indian sorts are so well differentiated from each other that some partial sterility results when they are hybridized. *Ames, 1939; EA*

RYE: *Secale cereale*

We come closer to understanding the history of rye than that of any other major cereal crop. Unknown before the Iron Age, it originated first as a grain-field weed in Asia Minor. Its ability to mature a kernel under less exacting conditions than wheat and barley led it gradually to win out over these crops in mountainous regions and on the northern edge of the wheat belts. The steps by which the wild plant first became the weed and the fashion by which the weed became big-seeded enough to be worth cultivating have yet to be worked out. *Schiemann, 1932*

SORGHUM (broomcorn, kaffir corn, kaoliang, Johnson grass, feterita, grain sorghums, etc.): *Sorghum* species

The genus *Sorghum* includes many wild and semiwild grasses, mostly African, a few Asiatic. Most of them are at least partially fertile with one another and a number of different species and hybrids have been brought into cultivation in Africa and in Asia, probably at different times and by different peoples. They are used for forage, for fuel, for building material and thatch, for grain, for syrup, and for brooms and brushes. Both in Asia and in Africa they have been cultivated since very early times. The United States Department of Agriculture took the lead in studying them seriously and was followed by England. At the present moment they are one of the most rapidly developing of the world's great crops, and their real development has scarcely begun. Their origin probably is tied up with early agricultural interchanges between India and Africa, something about which we know almost nothing. Sorghum would be difficult material to use for an analysis of this relationship because there are so many species and the differences between them are so slight. *Snowden, 1936; EA*

WHEATS: *Triticum vulgare* and several other species

Discussed in detail in Chapter 4. Our first wheats probably developed from a weed or semiweed wheat. In Neolithic times a true-breeding hybrid of these with a weedy quack grass produced emmer wheat, whose cultivation as a minor cereal is just now in its last flickering stages. Later, perhaps in the Bronze Age, further hybridization with *Aegilops squarrosa,* a weedy grass, produced the first of the hexaploid wheats. Meanwhile by similar steps such crops as the Persian wheats and the macaroni wheats had been evolved. Crosses with them produced our modern bread wheats. Kashmir, Turkestan, Syria, and Abyssinia were all actively concerned with the development of wheats. Some or all of them may have been among the centers in which various wheats originated, but as yet we have little proof of that. *Sears and MacFadden, 1946; Schiemann, 1932; EA*

Sugar

SUGAR CANE: *Saccharum officinarum*

The sugar canes and related weeds and wild grasses have in the last few decades been shown to form a gigantic polyploid complex of interrelated elements in southeastern Asia. It includes weeds, cultivated canes, wild canes, and grasses used for thatch, fiber, and perfume. Some of the weeds have certainly contributed to the origin of some of the cultivated strains; wild-growing canes once generally thought to be ancestral to the cultivated sugar canes have now been shown to have been derived from them, probably in post-human times. What the whole complex would have been like if man had not interfered, no one can say as yet, though new evidence is accumulating rapidly. *Parthasarathy, 1947; Venkatraman, 1938; EA*

Fibers and Oil Plants

COTTON: various species of *Gossypium*

Reviewed in Chapter 4, pages 66–71. The cultivated elements in the genus are treated as follows in the latest monograph, that by Hutchinson:

Old World cultivated species:

(1) *Gossypium arboreum.* Five races too poorly defined for more precise naming: the Burmese, the Indian, the Bengalese, the Sudanese, and the Chinese.

(2) *Gossypium herbaceum.* Grown in eastern Mediterranean, southeastern Europe, Asia Minor to Chinese Turkestan. In addition to the typical form there are the varieties: (a) *africanum,* grown in South Africa, and (b) *acerifolium,* grown in southern and eastern India and in Africa.

New World cultivated species:

(3) *Gossypium hirsutum.* Central America to the United States — "upland cotton." The variety (a) *punctatum* is wild and cultivated by natives in Central America and the Caribbean. The variety (b) *Marie-Galante* is grown in the southern Caribbean and South America.

(4) *Gossypium barbadense* is the widely distributed cotton of South America. It spread to the Antilles, whence it was introduced into South Carolina and Georgia, from which the sea-island cotton was selected. Crosses between Sea Island and perennial *barbadense* produced the commercial Egyptian cotton. *Gossypium barbadense brasiliense* was differentiated in eastern tropical South America, from which it has spread to Central America and elsewhere.

Hutchinson and Stephens believe that cotton was first domesticated as a fiber crop in the Indus valley in eastern India at a very early date. The role of African species in the origin of the cultivated cottons is an obscure one. It is possible that a weed cotton was taken to India from Africa before being domesticated as a fiber plant. The collapse of the brilliant Indus civilization in the third millennium before Christ isolated the previously contiguous cotton-growing areas (since irrigation works were abandoned and the area became desert again). This isolation led to the differentiation of different species and races of cultivated cottons in the Orient.

The Hutchinson and Stephens hypothesis of the origin of the tetraploid New World cottons has been reviewed in Chapter 4. On their hypothesis the transfer of these new cottons from Peru to Central America at an early date, with subsequent isolation between these two major centers, led to the differentiation of two different species of cultivated cottons in the New World. *Hutchinson, Silow, and Stephens, 1947*

FLAX: *Linum usitatissimum;* and WEED FLAX, *Linum angustifolium*

The history of both these ancient plants needs to be reviewed in the light of the modern evidence. Both are apparently tetraploids, while most of the wild species are diploids. *Linum angustifolium* is a common weed around the Mediterranean. *Linum usitatissimum* has been widely cultivated both for its oil (linseed) and its fiber (linen) since the most ancient times. One of its first uses may have been as a grain. Flaxseed cakes and linen thread are known from the early Neolithic remains. The weed may have come from the crop or the crop from the weed. Both may have originated independently and yet either or both may have acquired useful variability by hybridization with the other. The center of diversity for fiber flaxes is in Europe, for oil flaxes, in India and Afghanistan. *Schiemann, 1932; Ames, 1939*

HEMP: *Cannabis sativa*

Of very ancient use in the Orient, it is probably the oldest of our major fiber crops. Grown for fiber, oil, and drugs. It is a rampant weed on dump heaps, particularly when the soil is rich, and probably originated as a camp-follower weed. See pages 149–150. *Ames, 1939; EA*

OLIVE: *Olea europaea*

Apparently originated in Palestine where it has been used since ancient times. The wild form of the olive is widespread in the Mediterranean region and seems to be not merely the result of the cultivated form having run wild. *EA*

PEANUT: *Arachis hypogaea*

The peanut is known only in cultivation but related species are all South American; primitive forms of it have been recovered from ancient Peruvian tombs. It is therefore most certainly pre-Columbian in the New World. Yet up until the Peruvian excavations the experts were certain that it came from the Old World, so widely is it disseminated there, with every appearance of having been grown for a very long time in Asia and Africa. In fact, the old argument used to be whether it came from Africa or from Asia; so loath have botanists been to consider the hypothesis of pre-Columbian contacts across either the Pacific or the Atlantic that the evidence has never been reviewed to see whether or not it might be pre-Columbian in both hemispheres. One fact is already on record. The most primitive type of peanut, the same narrow little shoestrings which are found in the Peruvian tombs, are commonly grown today not in Peru, but in South China. How did they get there? If the Spanish or Portuguese carried them, a survey of the world's varieties of peanuts would produce some evidence of such a trip. If they went at some other time and by some other route the survey would produce evidence, for peanuts are a most varied crop. We already have enough opinions on record; what we need are more facts. *Ames, 1939; EA*

SESAME: *Sesamum orientale*

Americans are usually familiar with this crop only when the white nutty seeds are used in baked goods and Syrian candies, but it is one of the great oil crops of the world. Its wild ancestor is unknown. There are several species in Africa, but the crop is of tremendous antiquity in India. *Ames, 1939*

SUNFLOWER: *Helianthus annuus*

Reviewed in detail in Chapter 11, pages 186–206.

Forage Plants

ALFALFA: *Medicago sativa*

Originated anciently in the Persia–Asia Minor area. Natural hybridization between different species of Medicago certainly played a role in its development, though the exact importance has not been demonstrated. *EA*

BLUEGRASS: *Poa pratensis*

The common pasture and lawn bluegrasses are among the world's most complex mixtures of polyploid hybrids. From the cytological evidence it is probable that bluegrasses are a mixture of two or three wide crosses between distantly related species of *Poa,* themselves already polyploid. Bluegrass has the capacity to spread widely and rapidly with man's pasture animals and is known to have been introduced into the eastern United States in this fashion. (Kentucky bluegrass is not a true native of Kentucky any more than are Kentucky moonshiners. In the strict biological sense, both the moonshiners and the bluegrass are most probably transplanted Asiatics who came here by way of Europe.)

Bluegrass is certainly not native to all of Europe and may well be one of those plants which evolved along with cattle and horses and spread into Europe with these animals. The apparent diversity of bluegrasses in the Balkans suggests that bluegrasses may have been introduced there, when in a more nascent state, by early Neolithic herdsmen. *W. L. Brown, 1942; EA*

COWPEA: *Vigna sinensis*

The black-eyed pea of Texas and parts of the South. Of great antiquity in both Africa and India. Both a forage crop and a vegetable. *Ames, 1939; EA*

Vegetables

BEET, SUGAR BEET, SWISS CHARD: *Beta vulgaris*

Beets were certainly domesticated first as a leaf vegetable, then as a root crop, and finally as a source of sugar. They are probably derived from *Beta maritima,* a variable species of the Mediterranean region. Some of this variation may be due to natural hybridization with *Beta patula,* a very closely related species of Portugal and the Canary Islands. *Schiemann, 1932*

BROAD BEAN: *Vicia faba*

Though one of the world's commonest and most important beans, it does not take kindly to the climate of eastern North America, and is almost unknown to Americans. Its large flat brown bean is produced on upright, gray-green plants which are strikingly different in aspect from any other common vegetable. Widely grown since the Stone Age, it has centers of variation in both Abyssinia and Afghanistan. A closely related species is common in the Mediterranean and a wild-growing form has been collected in the Algerian mountains. *Ames, 1939; Schiemann, 1932; EA*

CABBAGE: *Brassica oleracea*

The cabbage vegetables (kohlrabi, cauliflower, kale, Brussels sprouts, etc.) are the European counterpart of the Asiatic mustards. They were originally grown for their oily seeds and, mostly in historic times, were gradually selected for their succulent leaves. They spring from *Brassica maritima,* a variable Mediterranean species with numerous local races and varieties. Since this is one of the few instances in which we can point to the exact complex from which a cultivated crop originated it would seem to be excellent material in which to investigate what happens when a plant is domesticated. *Schiemann, 1932*

CARROT: *Daucus carota*

Matskevitzh demonstrated that the diversity of the cultivated carrot in Afghanistan was way beyond anything known in Europe. The relationships of the various cultivated Asiatic forms to the European ones and to the widely distributed weed carrot, or Queen Anne's Lace, have never been worked out. *Matskevitzh, 1929*

CHICK-PEA: *Cicer arietinum*

Americans know this important legume, if at all, by the variety which is sold as a large irregular white dried pea, usually under the name of "garbanzo." It was commonly known to the Romans as *cicer,* and the orator Tully was known as Cicero because of a prominent wart resembling a chick-pea. Small, dark-colored, high-quality varieties are common in India. Not known as a wild plant. The genus is largely Asiatic, and it was presumably domesticated there. *Ames, 1939; EA*

CUCUMBER: *Cucumis*

Cultivated in India since ancient times. A semiweedy form of it is also found in that part of the world, and *Cucumis hardwickii,* native to the Himalayas, has frequently been cited as a possible ancestor. The relationships of these three entities have yet to be determined. *Ames, 1939; EA*

EGGPLANT: *Solanum melongena*

Apparently originated in India where there are related species and varieties which are weeds and semiweeds. In the Orient the crop varies in size, color, and shape from one variety to another and only a few of these types are common in Europe and America. *EA*

FENUGREEK: *Trigonella foenumgraecum*

This charming little plant which looks something like a dainty alfalfa with the flowers of a small sweet pea has been a camp follower and weed with man for a long time and has been put to minor uses. It was once a fairly important drug and the seeds are still used medicinally. In some Mediterranean countries it is planted as a forage crop, and in India it is used as a green vegetable and the seeds are used in curries. It would be difficult to say in what places it has just run wild and in what it is probably native. *Ames, 1939; EA*

GOURD: *Lagenaria siceraria*

The gourd is now universally admitted to have been cultivated in both the Old World and the New in pre-Columbian times. It furnishes dishes, water bottles, floats for nets and rafts and primitive life preservers, as well as food. There is great variation between the various cultivated varieties, the fruits of some sorts being of incredible size. A monographic study of its variation would be a monumental undertaking, but would yield unique information about the migrations of primitive man. There are no other species in the genus, which is thought to be native to the Old World. *Ames, 1939*

Jack bean and various other species of *Canavalia* beans:

Here belong the jack bean, the sword bean, and several lesser known species. They have been used since very early times in the Orient, in the West Indies, and in Polynesia. Until their classification has been more carefully studied, statements as to their probable origin and history have little scientific basis. *Burkill, 1935; EA*

Common or kidney bean: *Phaseolus vulgaris*

The origin of this important legume remains unknown. It is known to have been widely cultivated in the New World in pre-Columbian times. Its use in the Orient is widespread; no comprehensive survey of its varieties has even been made, and of our common plants it is one of the least known scientifically. *Ames, 1939; EA*

Lentil: *Lens esculenta*

An important element in the diet over much of the world, though little used in the United States. It has been grown since Neolithic times, and is thought to have originated in western Asia. *Ames, 1939*

Lima bean: *Phaseolus lunatus*

Authorities have never agreed as to whether the big Lima and the smaller, flatter, more pointed-podded Sieva Lima should be classified in one species or two. The former has its center of diversity in South America, the latter in Central America. Lima beans have been recovered from pre-Columbian graves in Peru in considerable quantity. *EA*

MUNG BEAN: *Phaseolus aureus*

One of the common pulses in India, where it has been used since ancient times. It is the species most commonly served as bean sprouts in Chinese restaurants. Apparently originated in India. *Ames, 1939*

THE MUSTARDS: several species and varieties of *Brassica, Eruca sativa*

The European who travels across northern India in the wintertime gains a new understanding of the continuing importance of these ancient crops. They are then in bloom, and the countryside is brightened with their yellow flowers. One can travel hundreds of miles and never be out of sight of them for a moment. The Brassicas are occasionally grown as a pure crop, though ordinarily in mixtures of two or more species and varieties. *Eruca sativa,* another mustard which produces an inferior grade of oil, does better on sandy lands and in dry seasons. It is frequently, though not always, grown separately. The mustards are grown for the oil in their seeds but are also used for greens and as a condiment, and I have seen them being chopped up for cattle fodder.

The mustards, whose use in the Orient is of great antiquity, give every indication of having originated as weeds which followed man around and were eventually put to various good uses. As fairly troublesome weeds they still have a much wider distribution than as useful crops. *Schiemann, 1932; EA*

PEA: *Pisum sativum*

The garden pea and the closely related field pea are known only as cultivated plants. Garden peas have been excavated at the Swiss lake-dwellings, a Stone-Age cavern in Hungary, and the site of ancient Troy. All of these were like modern peas, but much smaller. Both have two centers of variation, one in Afghanistan, and one in the Mediterranean. There are fairly closely related species of Pisum in the Mediterranean area. *Schiemann, 1932; Ames, 1939*

PIGEON PEA: *Cajanus Cajan*

The pigeon pea in great variety of plant type and kernel size is widely distributed through the tropics as a cultivated plant and a weed. It forms an important item of the diet in India and parts of Africa. All authorities suppose that it must have originated either in India or Africa, though no wild relative is known. Its origin is as much a puzzle as is that of *Zea mays. EA: Watt, 1885*

RADISHES: *Raphanus sativus* and several other species

One of our most ancient cultivated plants. It was already well established as a vegetable and an oil plant by classical Egyptian times. There is a related kind of radish, commonly classified as *Raphanus raphanistrum,* an ancient and ubiquitous weed in those portions of the world, such as California, with a Mediterranean climate. In the Orient, radishes play a much wider role than in the Western world. There, in addition to the uses we know, they are grown as stock food, for a winter vegetable which can be stored (a few of these types are also grown to a limited extent in Europe and North America), and for their seed pods, which are either eaten fresh or cooked as a green vegetable; specialized varieties with long narrow pods have been developed for this latter purpose.

Radishes give every indication of having originated from weedy plants which became camp followers and were put to various purposes. The exact relation between the various types of cultivated radish, the weed radish, and the ultimate prehuman elements in the genus, are not matters for discussion until we have more facts than are now available. On the face of it, the weed radish would seem to have given rise to the cultivated sorts. If so, when, where, and how did the weed radish develop? Methods are now available for finding the answers to such questions, but they are laborious and time-consuming. *Schiemann, 1932; EA*

RED PEPPERS: various species of *Capsicum*

The reference books will tell you that the source of chili powder, red pepper, sweet peppers, and so on, is a single species, *Capsicum annuum*. Recent work shows that at least five species are used, one of which is an Oriental weed, not yet reported from the New World. Peppers have certainly been cultivated in the New World since pre-Columbian times. They are widely used in the Orient, particularly in southeastern Asia. A thorough monographic study of their varieties on a world-wide scale would be of tremendous practical and theoretical importance. Until this is done we can know virtually nothing about their history.

Smith and Heiser classify them as follows: *Capsicum annuum:* the common sweet and hot pepper of the United States and Europe. Widely grown in the tropics. *Capsicum frutescens:* tabasco, an even hotter hot pepper. Widely distributed throughout tropical and subtropical America. Cultivated throughout the warmer parts of the world. Wild-growing in many Pacific Islands. *Capsicum pendulum:* extensively grown on the west coast of South America.

After a detailed description of the morphology of the first two, and a discussion of their crossing relationships (they are mostly pretty sterile with one another), the authors wisely conclude, "As soon as the taxonomy of the other species of this genus is understood and the distribution of the species worked out more accurately, the genus *Capsicum* should furnish useful material to those interested in the study of the origin and spread of cultivated plants in the Americas. However, with our present knowledge, speculation as to the place of origin of these species is out of the question." *Smith and Heiser, 1951; EA*

Scarlet runner bean: *Phaseolus multiflorus*

Rarely grown in Europe and the United States as a vegetable or as an ornamental climbing vine. A common vegetable in Central America, where numerous varieties are to be found in native markets. Artificial hybrids with the common bean are semifertile, but a limited amount of back-crossing is possible. This may have been responsible for some of the diversity in Central American beans. *EA*

Soybean: *Glycine max*

Of great antiquity in the Orient. The commonly cultivated sorts have a malformation of the stem and inflorescence which changes them from a trailing vine to a stiff upright plant. Varieties with the original trailing habit are still grown for forage and food in India and Java. The original type is reported to be still extant. *Ames, 1939; EA*

Spinach: *Spinacia oleracea*

Probably originated in Persia. Spread into Europe in the Middle Ages by way of Spain. Of greater antiquity in the Orient. *Ames, 1939*

SQUASHES AND PUMPKINS: various species of *Cucurbita*

Our winter and summer squashes, our pumpkins, and our bright-colored ornamental gourds until recently were supposed to belong to three species: *Cucurbita maxima, C. moschata,* and *C. pepo.* It has recently been shown that *C. mixta* is also involved as well as *C. andreana.* Bird's excavations in South America have demonstrated that the related perennial *Cucurbita ficifolia* was cultivated there in very early times. The wild and weed Cucurbitas of North and Central America need very careful study to determine the relation between them and these various cultivated sorts. Primitive varieties of winter squash are known from the same parts of China which have ancient popcorns and primitive peanuts. *Whitaker and Bohn, 1950; EA*

TOMATO: *Lycopersicon esculentum*

The cultivated tomato certainly springs from a genus of small-berried semicultivated weeds which are native to Peru. By some means as yet undetermined it acquired much greater size, and a more upright habit of growth. It spread to Mexico, apparently in pre-Columbian times, and was a more important element in the native diet there than in Peru. See pages 108–110. *Jenkins, Luckwill, 1943; EA*

TURNIP: *Brassica Rapa*

From the oil-seed Brassicas a number of other uses have eventually been evolved. The various Chinese cabbages spring from this complex as well as our common turnip. *Schiemann, 1932*

URD BEAN: *Phaseolus mungo*

Another of the common pulses of India, where it is grown in great variety and is an important element in the common food of the common people. Probably originated in India. *Ames, 1939*

Fruits

AVOCADO: *Persea americana*

One of the most important elements in the diet of Central America, where it is as much a butter substitute as a salad fruit. Apparently originated through a series of hybrids and back crosses between wild American species. Extensively cultivated in pre-Columbian times from Colombia to Mexico. See pages 102–106. *Anderson, 1950*

BANANA: *Musa sapientum*

Earnest attempts to classify the wild and cultivated bananas of the world have largely served to show what a complicated problem it is and how much more will have to be done before the origin and history of the group can be authoritatively summarized. The cultivated forms certainly originated in cultivation, some of them as hybrids, and the Malay Peninsula seems to be the chief center of origin. *Cheeseman* (Kew Bulletin, *various dates*); *Burkill, 1935*

CITRUS FRUITS (orange, lemon, grapefruit, kumquat, lime, sweet lime, citron, calamondin, etc.): *Citrus* species

This entire complex hybridizes readily and a number of natural and artificial hybrids are known. Southeastern Asia is apparently the home of the group, but it will take critical field and laboratory work to determine which elements in the complex are prehuman and which are post-human. Some of the so-called "wild" citrus fruits which have been collected in the tropics are either seedlings from cultivated sorts or actual remnants of native orchards which had reverted to brush. *Webber and Batchelor, 1943; EA*

COCONUT: *Cocos nucifera*

Perhaps, from a world viewpoint, our most important cultivated plant, a source of timber, thatch, beverage, oil, condiments, and fiber to many peoples, some of whom are almost wholly dependent upon it. Apparently domesticated very early in southeastern Asia, where there is a fossil record of the genus. Impressive documentary evidence that the coconut reached the New World in pre-Columbian times has recently been put on record. *Bruman, 1947*

DATE PALM: *Phoenix dactylifera*

Cultivated since very early times. The wild form is not known. Apparently originated in the region of western India and the Persian Gulf. *Popenoe, 1920*

FIG: *Ficus carica*

The cultivated fig has been known since very early times and appears to have been domesticated in southern Arabia. Three closely related species which are known to hybridize readily with one another are found in Persia, Mesopotamia, and Arabia; and it is in this part of the world that the cultivated fig is most variable. *Condit, 1947*

GRAPE: *Vitis vinifera*

Commonly used in the early Neolithic. Like many of our common European fruits it apparently spread out of the Transcaucasus–Turkestan area. Wild-growing forms are common there and are said to be extremely variable. They illustrate the kind of problem discussed in Chapter 6, and their relation to previous cycles of cultivation has yet to be determined. Artificial hybrids with American native species produced such varieties as the widely grown "Concord." *Schiemann, 1932; EA*

GUAVA: various species of *Psidium*

The cultivated guava was known from Mexico to Peru in pre-Columbian times. There are a number of wild or run-wild species of guava in the New World and they are characteristically weedy, coming up along roadsides, in abandoned clearings and the like. Determining the probable prehuman complection of this entire complex would require prolonged and intensive research. One species reached the old World at such an early date that it is frequently referred to as the "Chinese guava." *Popenoe, 1920*

JAPANESE PERSIMMON: *Diospyros kaki*

Originally from China. Hume believes it to be a hybrid derivative of two or more wild species. *Popenoe, 1920*

JUJUBE: *Zizyphus*

Has been cultivated in China for at least 4000 years. At least two species are involved. *Popenoe, 1920*

MANGO: *Mangifera indica*

One of the oldest and most important of the tropical fruits. It is an important element in the diet of southeastern Asia, where it undoubtedly originated, probably as a polyploid hybrid between species as yet undetermined. *EA*

MELON: *Cucumis melo*

Cultivated since ancient times. It is one of several crops indicating a very ancient agricultural interchange between Africa and India. The genus as a whole is mostly made up of African species; the cultivated melon is more variable in western Asia than in Africa. Some authorities have supposed it to have originated in Africa, others consider Asia more probable. *Ames, 1939*

MOMBIN: various species of *Spondias*

Various forms and species of these tropical fruits have been grown in the New World, apparently since very ancient times. Until the whole group is carefully monographed we shall have little basis for suggesting what the genus was like in prehuman times or for indicating the routes and the times at which the cultivated species spread. *EA*

PAPAYA: *Carica papaya*

The common breakfast fruit of most of the tropics. Widespread in the tropics where it frequently becomes a weed. The related species and genera are all in the New World and it probably was domesticated in the Amazonian basin. It tastes rather like a slightly butterier muskmelon, but has the surprising capacity of being able to digest meat. A tiny sliver of papaya laid on a steak will liquefy the meat all along the zone of contact. For this reason the dried coagulated sap, imported under the name of *papain,* has become increasingly important. It was originally used in patent medicines for dyspepsia, but modern technology finds it useful for such purposes as preventing cloudiness in bottled beers, and it has become an important item in world trade. *Tainter* et al., *1951*

PEACH: *Prunus persica*

The peach is so thoroughly trained to man's ways that it runs wild spontaneously along roadsides and in rocky pastures through many parts of the world, including the United States. This fact makes it impossible, without more initial evidence, to interpret the wild-growing peaches of Kashmir and western China. The peach is known to be of great antiquity in China, where many varieties were in cultivation before 2000 B.C. *Schiemann, 1932; EA*

PINEAPPLE: *Ananas comosus*

The cultivated pineapple is not unknown in the wild, although all the other members of the genus, and for that matter practically the whole family, are native to the New World. The cultivation of the pineapple was widespread in South and Central America by the time of Columbus, and it is presumed to have originated in South America. *Collins, 1949*

Pome fruits

APPLE: *Malus pumila*

All the pome fruits (that is to say, all fruits with a true core, such as apples, pears, loquats, etc.), the wild pomes as well as the cultivated sorts, are a gigantic polyploid complex. It includes many hybrids, some of which seem to be true-breeding, because the apparent seeds are really just a budding off from the mother plant, and not truly sexual offspring. So-called wild species of apples grow in various parts of Asia in great variety. Many of them have been introduced into the Western world as ornamental crab apples, and the classification of these cultivated ornamentals, the wild-growing Asiatics, and the cultivated kinds of Occidental and Oriental orchards is a perfect mare's nest of a problem. These wild and semiwild apples of Asia are as complex a technical problem as are their pomaceous cousins, the thousands of meadow and woodland hawthorns which bedevil American naturalists with their infinite variety.

It is likely that a relatively few prehuman species of Malus found in man's upsetting of the natural vegetation a glorious evolutionary opportunity. The hundreds of forms of so-called wild apples in various parts of Asia are probably for the most part weeds, camp followers, and encouraged semidomesticates which man has unconsciously been breeding in that part of the world since very early times. The cultivated apple is thus one small segment of a much larger and more intricate problem. The so-called wild apple of Europe, *Malus sylvestris,* is certainly one of the elements from which our cultivated European varieties have been bred. It is common throughout much of the Balkans, and though I saw thousands of trees there, not one of them looked as if it might be part of the truly indigenous flora. All I saw were growing in old pastures, or near farm buildings, or along old pathways. Most likely they are remnants of a primitive cultivated fruit which in Neolithic times was brought into Europe from its ultimate beginnings in the Caucasus. *Schiemann, 1932; EA*

Pear: *Pyrus communis*

The story of the cultivated pear is much like that of the apple except that it is a simpler one; there is little or no polyploidy, and fewer original wild species were involved. Pears closely related to the cultivated sorts are "wild" from Turkestan and the Caucasus to central Europe. Natural hybridization is apparently common. The so-called wild species and subspecies were certainly affected by man; the story remains to be worked out. *Schiemann, 1932; EA*

Quince: *Cydonia oblonga*

Wild-growing quinces are known in Asia and Asia Minor. In part these certainly represent cultivated quinces which have run wild. All we can say for certain is that the quince is pre-European. *Schiemann, 1932*

Pomegranate: *Punica granatum*

Apparently native to Persia, it is known to have spread from China to the Mediterranean by the beginning of the Christian Era. *Popenoe, 1920*

PLUM: common plum, *Prunus domestica;* sloe, *Prunus spinosa;* Balkan plum, *Prunus cerasifera*

It has been shown cytologically that the common plum is a hexaploid, and that its parents are very probably the sloe and the Balkan plum. The sloe, a spiny little shrub with a tiny bitter dark-blue fruit, is common in European hedgerows, and is one of those semidomesticated plants which have learned to profit by man's upsetting of the natural balance of things. *Prunus cerasifera* is the source of the fiery plum brandy (*slibowitz*) produced by Balkan peasants. Though the trees are not usually deliberately planted, they are universally encouraged. These facts suggest that two primitive semidomesticates, the sloe and the Balkan plum, produced a polyploid offspring under man's influence as early as the Neolithic. This early domesticate spawned a semiweedy shrub (*Prunus insititia,* the ancestor of our damson plums) which is wild-growing from the Caspian to south and middle Europe. *Schiemann, 1932; EA*

STRAWBERRY: *Fragaria grandiflora*

The cultivated strawberry arose in the eighteenth century as a true-breeding polyploid hybrid from artificial crosses between the wild Chilean and the wild American strawberries. Previous to this time various other American and European species had been grown to a slight extent. This vigorous new polyploid strain quickly became an important fruit crop both in the Old World and the New. It is the one crop of world importance to have originated in modern times, though the new American blueberry hybrids may someday be equally important. *Schiemann, 1932; EA*

WATERMELON: *Citrullus vulgaris*

Of ancient origin, most certainly in Africa where bitter-fleshed sorts are still found in the Kalahari desert region. When the history and distribution of the various races of watermelon are known, the early appearance of watermelons among the Indians of the Southwest should be investigated to determine the exact route by which they came. Africa–Moors–Spain–Mexico is the most probable route, but this has not yet been documented. *Ames, 1939; EA*

X I

Sunflowers—The One Native American Crop

EVER SINCE I FIRST SAW acres of wild sunflowers blossoming in gold and green confusion all down the back slopes of Dago Hill in Saint Louis, I have been intoxicated with sunflowers. They happen to be of key significance in the story of plants and man; if they were not I should be tempted to tell you about them anyway. I wish it were possible fully to communicate my enthusiasm for these lusty brilliant flowers, these coarse and resinous weeds which have been so closely tied up with man during his varied history on this continent. Crop plant, garden flower, roadside weed, growing unasked but seldom unwanted on vacant lots and rights of way. I wish one could in a few sentences paint the full pageant of them in any one year. The colonies of them in railroad yards in our larger eastern cities, the thousands upon thousands of acres in the Great Plains, long golden rows of them bordering roadsides and drainage ditches throughout the West, regular lines of them planted as windbreaks in some of the irrigated valleys, little bunches of them in neat Hopi gardens below the mesa tops, acre after acre grown for oil in western Canada. Or the fields of them in Peru and in Europe, single rows of them in allotment gardens at the fringes of cities, masses of them in Hungary and the Ukraine where they are grown for cooking oil. But with all my enthusiasm for them, with all my understanding of their high yield, I do not advise you to try growing them on an American farm year after year in efficient mass production.

No world crop originated in the area of its modern commercial importance and sunflowers are no exception. They originated here and their great commercial area is in southwestern Europe. We grow wheat, maize, and soybeans in the area where the sunflowers are native, wheat from the back corners of Asia and Asia Minor, maize from the tropics, and soybeans from the Orient. Even in the Orient the same principle holds for the soybean. Its great commercial area is to the north in Manchuria though it is a native of southeastern Asia.

With many tropical crops the center of production is even in the other hemisphere from the center of origin. Coffee originated in the Old World and is grown in the New. Chocolate originated in the New and is grown in the Old. Rubber and quinine went to Malaysia from the New World; bananas and citrus fruits came here from there. The underlying reason for this shift is not just that a prophet is not appreciated in his own country. In the region where a crop was domesticated there are the maximum number of pests and diseases which have been evolved to prey upon that particular kind of plant. As soon as the crop is grown in large numbers, when whole fields, or in these days of mass production, whole landscapes, are covered with one kind of plant, protecting it from the various insects and fungi which feed upon it becomes increasingly difficult. The farther you get from its center of origin the more of its pests can you hope to leave behind. This seems to be the main reason why the cultivated sunflowers have never become the important oil crop in the United States which a number of enthusiasts have hoped they might be.

If you want to see for yourself why it would be difficult to make sunflowers a permanent commercial possibility in most of this country, go to the nearest railroad yard or dump heap where the small-headed "wild" sunflowers are growing and harvest twenty-five or fifty plants at random. Bring them back home and note the numbers and kinds of insects which are using the plant for a lunch counter. Insects tunnel back and forth in the thick white pith of the stem, insects chew the coarse, heavy leaves and the moplike rootlets, insects eat out the insides of the seeds. Many of the seeds look all right from the outside; open them up and you find that some beetle or moth or fly, having fed on the contents, is now comfortably curled up and is all ready to use the empty shell as a winter home.

Every so often someone learns that sunflowers grow well in the Middle West, finds out that they are high-yielding, and that the oil in their seeds is the finest table oil which can readily be grown in the temperate zone, and starts out to grow them as a crop. The story is nearly always the same. The first year they do well; sometimes even the second or third season. Then comes failure. Not only are sunflowers native to the United States, the whole genus *Helianthus* is centered here. The world's largest reservoir of sunflower insects and diseases is always close at hand. A few of these move into the lush pasturage of the sunflower field the first year and some of them find field conditions very much to their liking. They raise large families and next year there is a larger potential of sunflower pests to winter over and be ready for another year's sunflowers. Often by the third or fourth year there is not merely a bad infestation; there is virtually no crop at all.

The very facts which doom sunflowers to commercial failure in the United States hold out promise for a unique role for them in scientific research. They are the only world crop domesticated within the boundaries of the present United States. All their possible wild ancestors, suspected or unsuspected, are here. With a modest research budget one could bring together sunflower material for studying one of the major problems in biology: What happens to a plant when it is domesticated? What kinds of changes mark it off from the genuinely wild plants from which ultimately it was derived? What are the relations of these wildlings and the various weeds and semiweeds? Did the crop grow out of the weed, or the weed grow out of the crop? If they had separate origins how did they affect each other if at all? We now have the techniques, given authentic living material, to answer these and related questions with actual facts rather than with reassuring theories. These are questions of far-reaching importance to technology, to biological theory, and to social history. It is a more difficult field in which to get precise information than most scientists have realized. Many of them would begin to plan their research on the assumption that the basic taxonomic facts now on record are accurate and complete. As we have seen these are often misleading, frequently incorrect, and nearly always incomplete.

Given an understanding of what constitutes really critical data this is still an almost impossible field in which to work effectively. The practical job of getting the living materials to work with and growing them for several generations presents unseen difficulties. Our common north European crops are biologically speaking mostly Asiatic species like us; for authentic living material of their wild and semiwild precursors we must go to the back corners of the old world, to Anatolia, Turkestan, Abyssinia and the Caucasus. Our New World crops, aside from sunflowers, came up from South and Central America, and their precursors are tropical. It is easier to get to highland Guatemala or even to the Amazon basin than to the Caucasus or Turkestan in these days, but with living material from the tropics, getting the seeds together is only the first part of the problem. They are tropical plants and tropical plants are not at home in the Temperate Zone.

Temperature is only one of the differences; there are such complications as responses to day length. Suppose you set out to study the common bean and bring up a collection of Central American and South American beans for your work. Plant them in your garden and a considerable number of them will just grow and grow and grow, nothing but leaves, waiting for the days to shorten. With sunflowers most of these practical difficulties are avoided, though even with them some of the southwestern varieties need careful handling if they are grown farther north.

With sunflowers all the possible beginnings of the crop are here unless some of them may have passed out of existence. The genus *Helianthus,* to which all species of sunflower belong, is an American genus and most of its species are native within the boundaries of the United States. The annual sunflowers, a separate group within this genus, are all native to the United States and it was somewhere within our borders that most of the process of domestication went on. This means that the botanical, anthropological, ethnological and historical data required to block out a working hypothesis are all at hand. Once set up such a theory and the living materials on which it can be tested and verified or rejected are all to be found within the United States or northern Mexico.

Fortunately a few American scientists have already seen the special opportunities presented by our sunflowers. Volney Jones, the ethnobotanist, has brought together all existing ethnological and archaeological collections, seeds of varieties from all existing Indian tribes which grow sunflowers, excavated seeds and heads from ancient burials and dump heaps. Dr. Charles Heiser has started in on the long, complicated technical job of finding out what kinds of sunflowers there are, weed, wild, and cultivated, and how they are related to each other. When he began his work five years ago, only the barest outline of the facts had been published and of that little on record a good deal was incorrect.

Look in the reference books under cultivated sunflowers and you will be informed that there is one species, *Helianthus annuus,* and that it is native to the Great Plains. They will add that it has been brought into cultivation and that it has spread into waste places in other parts of the country. Persist in your search through the botanical literature and you will find that distinctive names have been proposed for the common sunflower which is native to the Great Plains. Talk to Dr. Heiser about them and he will start telling you about this race and that race, this kind of weed, that cultivated sort. Nor will he stop at the boundaries of *Helianthus annuus.* He will go right on and insist that we must also consider *Helianthus petiolaris* and *Helianthus cucumerifolius* and *Helianthus bolanderi* at the very least. This is typical of what happens when we start looking into an apparently simple problem in the origin of cultivated plants. The easy generalizations which get into the encyclopedias are only a crude approximation of what we find when we go into a problem carefully.

To follow Dr. Heiser even approximately, we shall have to consider, at the very least, twelve different races, species, and varieties of sunflowers. This is too many for the uninitiated mind when all twelve are sunflowers and more or less alike anyway. Talking about *Helianthus* this and *Helianthus* that, is like jumping into a Russian novel full of Olga Stefanovna's and Maria Alexandrovna's and trying to keep the characters in mind until the story gets under way. The last time I read *War and Peace,* the publisher had thoughtfully provided a bookmark on which the entire cast were listed by families with their full names, their intimate family names, and even their nicknames. With that as a guide I went through the two volumes swimmingly; perhaps we can use something of that sort here. First then:

The Cast of Characters

Helianthus annuus
- Cultivated sorts
    - Modern commercial varieties, Mammoth Russian, etc.
    - Ornamental varieties
    - Hopi Sunflower
    - Other Indian varieties
    - Northern Mexican Sunflower
- Weeds and semiweeds
    - Camp-follower race
    - Great Plains race
    - Southwestern race
    - California Valley race
    - East Texas race

Helianthus petiolaris
Helianthus cucumerifolius — Texas
Helianthus argophyllus — Texas
Helianthus bolanderi — Great Valley of California
Helianthus exilis — Serpentine endemic, California

Of the cultivated varieties all except some of the ornamental sorts have one large head which is at the summit of the plant. The stem is unbranched, the coarse heart-shaped leaves extend out a foot or more from the stem. Towards the top of the plant they gradually become smaller and are succeeded by little resinous bracts which are shingled thickly one layer after another all over the back of the flower head. The actual flower is a small affair, like a little dark-red tiger lily less than a quarter of an inch across. Hundreds of them are set close to one another in spiraling mathematical regularity all over the face of the bloom. Surrounding them are the specialized ray flowers, each with its corolla grown out lopsidedly into a long yellow petal-like ray. The flow-

Groß Indianisch Sonnenblum. Flos Solis Peruuianus.

FIGURE 16. Sunflower. From the herbal of Mattiolus, 1586

ers begin to open around the rim and proceed inwards in regular succession. Early in the season one sees the sunflower as a central disk still dark with the glistening resinous bracts which protect the buds and around them an outer circle of the flowers which have opened, all crowded and polleny. As the season advances the last of the flowers to open in each head will make a bright little red-brown tuft in the center of the disk surrounded by the enlarging green plaque of flowers which are ripening into seed.

This huge head of bloom really does turn with the sun. As soon as the ray flowers are conspicuous and before any of the flowers have actually opened, it faces the sun early each morning and turns slowly with it all during the day. Before sunrise the next morning it will still be facing the west but by half an hour later it will have swung back all the way towards the east. This movement is repeated day after day until the head is nearly through flowering, and is heavy with developing seeds.

The big conspicuous heads of the cultivated sorts are a freak condition, a simple inherited abnormality like double flowers, or sweet kernels, or any of the other strange departures from the normal that man puts to his uses. It would be more exact to describe these single-headed varieties as being unable to branch. Compare one of them with a rank, weedy sunflower from the railroad track. The two plants are almost identical aside from the size of the flower head and the degree of branching. The weed sunflower has a main stem with a flower head at the summit, just like the cultivated sorts, but at every leaf on the main stem there is a branch which shoots out vigorously with a flower head at the end of it, and at each of its leaves there are branches which may themselves be branched and so on up to the limit of what one set of roots can feed. A large-headed sunflower is made by essentially the same process as the big shaggy heads of chrysanthemums which are sold at football games. In each, by preventing branching, all the strength is forced into one head. With the chrysanthemum this is done by the florist, who pinches out all the side buds when they first develop. The sunflower does it internally. When we cross normal and single-headed plants, we find that single-headedness is recessive and that it is inherited in the simplest kind of way so that we can convert any branching sunflower to the single-headed condition by making the appropriate crosses.

So much for the big golden disks of the cultivated sunflowers. Where did they come from? This is quite a question. There are branched annual sunflowers growing without the aid of man all the way from the vacant lots of suburban Los Angeles to the railroad yards of New York City and Long Island. It requires only a slight familiarity with sunflowers to see that though they may all be *Helianthus annuus* they certainly belong to different races of that species. They vary in size, in blooming period, in leaf shape and texture, in the proportions of the flower head, and in the color and shape of the ray flowers, some broad and orange-yellow, others narrow and lemon-yellow. Nor are these differences merely due to the different environments. When Dr. Heiser grows samples of many strains in his experimental plot at the University of Indiana, most of the differences persist and some of them are even accentuated.

If we are going to try to make some sense out of these hordes of weeds and wildlings, central Iowa is a simple place to begin. We will find a lot of sunflowers there, though not as many as in Kansas, and they clearly are growing in two different places, in the city and on the farm. The rural ones grow much like those farther west in Nebraska and Kansas. They are found as a weed in some cornfields, along fencerows between fields, at thin places in overgrazed pastures, in deserted fields, in vegetable gardens which have been let run to weeds, and along railroad tracks and highways. Their city cousins grow in no such places. They are in the Des Moines railroad yards in the heart of town, on dumps along the river, and blossoming between ashes and rubble in the storage yards behind factories.

According to the books these are both *Helianthus annuus,* but if we collect twenty flower heads of each and lay them out on the table (the floor is even better so that you can look down on the whole lot) we find a tenuous sort of difference between the two sorts. Those from the country are smaller, with fewer, more slender, more lemon-yellow rays. Those from the city are larger and coarser with broader, more orange-yellow rays. More frequently the city sunflowers show a central zone of deeper, darker color, a so-called corona. If we have selected twenty-five from the center of the city and twenty-five from a cornfield well outside of town, though they vary a great deal in each lot, none of the city twenty-five will be quite like any of the country twenty-five. If we pick a set from a dump heap in back of a garage at the edge of town, on the whole this lot will look like the city set, but a few of the plants will resemble the country ones, and one or two may be indistinguishable.

These are two of the main races of *Helianthus annuus;* Dr. Heiser has not yet started designating them with Latin names so we may follow him in referring to them for the time being as Great Plains *annuus* and the Camp-follower *annuus,* two of the leading names, you will remember, in our cast of characters. "Camp follower" is a perfect description for the sunflowers found in Des Moines and many of our other big cities. They are one of the sights of Saint Louis, though most Saint Louisans, having seen them since childhood, take them very much for granted. They grow in great profusion in our railroad yards in Mill Creek Valley, the industrial trough which cuts Saint Louis roughly in half. They come up between the stones in the revetments along the levee, they grow on cinders, amid rubble and scrap metal. On eroded slopes in the industrial district they form pure stands, sometimes acres in extent and during the season when they are in flower they are a glorious mass of dark-green leaves and golden disks. They have a way of appearing in the most unlikely places, in pure clay as hard as a tennis court which was dug out in erecting a large advertising signboard, at the edges of ash pits in the alleyways, particularly in the poorer parts of the city. One good-sized plant was growing on the south side of a shoe factory, having somehow sprouted in the crack between the building and the sidewalk. There it was in the full blazing sun of a Saint Louis summer, five feet tall, and with several heads of flowers! A trip around the city in September will tell us a good deal about them, particularly if we notice the places where the sunflowers are *not* growing. They must have sun; a small signboard will cast enough shade to discourage them effectively in an otherwise suitable spot. The one thing they seem unable to tolerate is competition from other kinds of plants, particularly grass. At the edge of a cinder dump where the cinders get thinner and the grass begins to come in along the edges, the sunflowers are dwarfed by even a thin cover of weed grasses. When rubble and metal scrap are dumped in the middle of a comparatively fertile field, sunflowers grow in among the rubble where the grass cannot get a start; at its edge they get progressively smaller and in the thick turf of the rest of the field there are no sunflowers at all.

It is not that sunflowers do not appreciate rich soil. Take any of these weed sorts and grow them in a fertile experimental plot and they will grow up into husky giants, higher and more branched, with more and larger flowers. They are on the poorest soil because only there can they get away from the grass. There is something about grass which they just cannot tolerate. I once planted a series of flowerpots with increasing numbers of sunflowers to study their competition with each other and then another set with the same number of sunflowers in each pot but with increasing numbers of rye plants, rye being a vigorous grass comparatively simple to grow in an ordinary greenhouse. With increasing competition of their own kind they germinated just as well, grew just as well for several weeks, and then showed increasingly the effect of competition. With the pots of rye it was quite another story. As the rye was thicker and thicker, from pot to pot of the series, they germinated less and less well; they grew more and more slowly. One rye plant had a depressing effect upon four sunflower plants in a large pot from the very first week of its growth.

No, these weed sunflowers seem to have been bred as a camp follower. They are only happy when clothing some of the most repulsive scars man makes in the landscape. They are unknown in the country, rare in towns and villages, and common in cities. Within the cities, it is the slums, the alleyways, the dump heaps, the raw newly excavated sites where they are most at home. There may be some part of the sunburnt West with salty or sunburnt open soil where they were growing before the advent of man, but in all the places I have seen them, they were as much the result of his efforts as tin cans and trolley cars. When and how were they bred in the first place?

Before we can discuss this question intelligently we shall have to introduce *Helianthus petiolaris,* another member of our cast of characters. It is found in the freight yards in Saint Louis, or occasionally in other parts of the city, but its real home is farther west where it covers whole landscapes with yellow in the early summer. It is a much shorter plant than the common sunflower, frequently only knee-high, and it looks rather like a field daisy trying to be a sunflower and not succeeding too well. Technically it is distinguished from *Helianthus annuus* by tiny white tips, microscopically lacy, on the sheltering bracts around the seeds. There are so many of these little stiff white plumes that in the crowded flower head they show to the naked eye as a gray-white center to the dark disk. On the Great Plains they grow in much the same sort of places as the Great Plains *annuus;* in Saint Louis they are confined to drier and gravellier sites than most of those chosen by Camp-follower *annuus.*

Both sorts of *annuus* differ from *petiolaris* quite as much by their season as their size. Characteristically *annuus* does not flower until mid-summer and is not at its best until late in August. *Petiolaris* flowers with the first really hot weather. Early June always finds a few of them opening their flowers in the Saint Louis freight yards. Since both *annuus* and *petiolaris* grow in Kansas in great profusion, I have often wondered if Kansas Republicans realize what a thorny technical problem they have on their hands. If they have sunflowers for their emblem at a June convention it must surely be *petiolaris;* if the convention is in August it is probably *annuus,* particularly since, being Republicans, they will pick out the largest ones then in flower rather than the smallest. And as for a July convention! The problem is too technical; we shall have to call in Dr. Heiser.

He tells us that the relationship is a complicated one. He finds that the two species can be crossed, though with some difficulty, and that they produce vigorous hybrids intermediate between the two species. These vigorous intermediates, like many hybrids, are partially sterile, but it is possible to cross them back to *annuus* and to *petiolaris* and obtain three-quarter bloods. The three-quarters *annuus,* on the whole, look like *annuus,* although among them are some seedlings which give fairly clear indications of their mongrel ancestry. If these in turn are crossed back to *annuus* a second time, producing mongrels which are roughly seven-eighths *annuus* and one-eighth *petiolaris,* they give little or no hint of their exotic pedigree. If collected in Kansas they would unhesitatingly have been identified as perfectly good *Helianthus annuus* by any botanist, be he ever so expert. The only indication of partial hybridity is that on the whole they are a little smaller, and a little more slender, and a little earlier to flower than uncorrupted *annuus*. Therefore if the Kansas Republicans attend a July convention, or even one in late June, and being Republicans choose the largest sunflowers then in bloom as their symbol, they have probably picked out mongrel sunflowers of very dubious ancestry.

Whatever the Kansas Republicans may or may not think, it is a fascinating and intricate problem, this question of the Kansas sunflowers. Dr. Heiser has not yet gotten to the bottom of it, but it is clear that there is an active interchange going on between the two species. Having produced the first generation hybrids artificially as well as the three-quarters and seven-eighths mongrels he is now equipped to look intelligently at the hordes of yellow sunflowers between eastern Kansas and the Rocky Mountains. First-generation hybrids are not common, but they can be found, and sometimes patches of mongrels which could be referred to as a hybrid swarm. Much more common, however, are *petiolaris* which now give him every indication of having a little *annuus* in them, or *annuus* which match the seven-eighths-blood mongrels with *petiolaris*. It is these mongrels which are particularly to be found where man has greatly disturbed the habitat, at points where a state road has been shifted two or three times and various ditches and burrow pits have been dug. So extensive are these signs of a little filtering through of one species into the other, that Dr. Heiser is not yet prepared to say, or even to guess, what Great Plains *annuus* was like before man came along, or even to be certain that it was there in such quantities as it is today. Of one thing he is certain, that *petiolaris* has played an important role in the development of Great Plains *annuus*.

One of Dr. Heiser's first investigations did not concern the main stream of *Helianthus annuus* but it is interesting because it demonstrates so effectively the relation between man and the weeds that he unconsciously breeds, not only in *Helianthus* but in many other genera. This illuminating and entertaining side show took place in California (like many other side shows). When he began his work there were known to be three annual sunflowers in California. (1) *Helianthus annuus,* planted as a windbreak and an ornamental, and frequent as a weed on vacant lots, along irrigation ditches and the like. (2) A coarse woolly native species, *Helianthus bolanderi,* spreading as a weed among the ranches of the northern part of the Great Valley of California, and (3) *Helianthus exilis,* a strange little yellow daisy, something that you or I would not know was a *Helianthus* the first time we met it, which is confined to highly localized serpentine barrens in the Great Valley. What Heiser detected and then proved conclusively was that originally there had been only *exilis* and *annuus.* The two hybridized and though the hybrid was pretty sterile, enough backcrosses to *exilis* took place so that some of the weedy adaptability of *annuus* was carried over into *exilis,* producing the large weedy strain of it now known officially as *Helianthus bolanderi.* This in turn hybridized and rehybridized with *annuus,* perhaps with both weed and cultivated strains, and from these various and varying mongrels there was bred, under the stern discipline of natural selection, a set of superweeds combining the adaptability of *exilis* to California conditions and the suitability for man's strange upsetting of the natural balance of things, which is so characteristic of *Helianthus annuus.*

Just when this took place, Dr. Heiser does not yet know. He is certain it is a post-human phenomenon but whether it began in aboriginal times or not until after the arrival of the Spanish he is not yet prepared to say. For understanding the interrelations of man, weeds, and cultivated plants, it is a key experiment. It shows that a wide-ranging weed may undergo new and important developments far from its original home and that a species superficially so different that no ordinary person would think it was in the same group of plants may contribute some of the adaptability for new conditions.

Right now Dr. Heiser is working in Texas and uncovering a similar, though more complicated, state of affairs. In Texas there are two native annual sunflowers: *Helianthus cucumerifolius,* a low, dainty, much-branched species with ruffled leaves, and *H. argophyllus,* very much like the common sunflower, but covered with a handsome thick felt of silvery white hairs. He is finding that both of these hybridize with *annuus* and that practically all the *annuus* in Texas and much of that in Oklahoma has been perceptibly modified by the incorporation of a little influence from these two Texas species.

So much for a rough picture of the main elements from which our cultivated sunflowers must have sprung. There are still other sunflowers which might be considered. There is a variety or species in Florida much like the *cucumerifolius* from Texas; there is a little-known species in the Southwest; and there are various weedy strains of *annuus* which do not quite match up with either Great Plains *annuus* or the camp-follower race. One of the complications in the problem is the peculiar varieties grown by the Hopi Indians and by northern Mexicans. These have the single-headed character of most cultivated sorts but they are tall, slender plants with proportionately small heads, darker leaves, and with slender dark-purple seeds. They are longer seasoned than other cultivated sorts and the Mexican variety is even later to bloom than the Hopi.

From archaeological investigations we know that the history of these crop plants is a long one in the New World; we will know even more when Volney Jones gets all his evidence expertly fitted together. All that we can say now is that sunflowers were an important crop here in very early times. Sunflower heads and seeds, similar to some of the weed types, have been excavated from various pre-Columbian sites, some of them quite early. The sunflower had already a long and complicated history behind it when the white man first came to these shores.

From the Southwest we not only have sunflowers from excavations but we know they are most important among the Hopi not only as a food but as a purple body paint and in religious ceremonies. Such uses ordinarily indicate considerable antiquity, particularly with a conservative people like the Hopi. In the eighteen-nineties Alexander Stephen lived among the Hopi and kept a detailed journal in which he recorded the daily life of the people, particularly their various ceremonies. There are several mentions of sunflowers. At one ceremony he noted, "Just as this ceremony began, eight young men with whitened bodies came in on a run. Before daybreak this morning these eight young men whitened their bodies in Chief Kiva and naked save for breech cloth and silver belt they started off at a brisk trot. They gathered such fruit from the gardens as was approaching ripeness, some melons about the size of a cricket ball, chili beans, cockscomb, squash blossoms, and melon and other vines. With these they decked their persons, putting sunflowers in their hair and came back to Walpi on a swift run." In another ceremony the sunflowers were first worn and then were laid so as to form a continuous line of rosettes along the ceremonial trail of pollen. Stephen published sketches of the designs which had been painted on the walls of the kivas, or sacred lodges. Among these designs were realistic drawings of sunflowers of both large-headed and small-headed sorts.

In his classic excavations in the Southwest, S. J. Guernsey discovered a beautiful set of carved and painted wooden sunflowers which had been made in prehistoric times. Over thirty of them were packed in a big corrugated olla which was excavated at

Marsh Pass. Some of these are now on public display at the Peabody Museum. The sunflowers are of the small-headed sort, and similar in size and shape to those which can still be found growing wild in that region.

This is most of what is definitely known about the history of the sunflower. To get beyond this point we shall have to make estimates from known facts; we shall have to put two and two together; we shall have to use our imaginations. The general picture will probably be pretty close to the truth; in some of the details it will certainly be incorrect. What was our cast of characters doing before man appeared? *Exilis* was certainly a curious little plant restricted to the serpentine areas of California. *Argophyllus* and *cucumerifolius* also had very restricted distributions on sandy ridges in Texas, and along the coastal sands. *Petiolaris* was then a well-behaved species with little variation in size and form. It was like the smallest, daintiest, most early flowering extremes one now finds in the Great Plains. It was native then on sandy places; where there were natural blowouts it sometimes made sheets of yellow but it did not cover whole landscapes as it frequently does today.

But where was *annuus?* Well, certainly not in any railroad yards or dump heaps; these favorite sites were yet unknown. My guess is that it grew in those muddy little basins of Arizona and New Mexico where water collects after a rain. It already had the dark-green leaves of plants characteristic of such places, but they were smaller than most of those one sees today. There may possibly have been another desert-loving sunflower somewhere else in the Southwest but if so it, too, was restricted to one characteristic habitat.

When seed gatherers first moved into the West, the oily seeds of the sunflowers, with their brilliant purple dye, were among their most natural prizes. They were gathered and stored, they may occasionally have come up in dump heaps, the variability of the species increased a little bit with the mixing of different strains. Finally, through man's interference this proto-*annuus* and *petiolaris* were brought together and since all sunflowers are naturally cross-fertilized, once they came together their chances of being crossed were excellent. Out of a series of such hybridizations a camp-follower weed was bred and generation after generation it became more skillful at fitting into the strange new niches man makes in the natural scheme of things. Among these mongrel weeds hundreds and perhaps thousands of years later, there turned up freak unbranched forms. From them the domesticated single-headed sunflowers have descended. They were passed on to other tribes and other groups of tribes; they traveled north and south and east. Meanwhile *annuus* and *petiolaris* meeting more and more frequently in the Great Plains bred a mongrel recombination, Great Plains *annuus,* which was at home over wide areas in the West, particularly in disturbed situations.

By the time the white man arrived, the camp-follower weed, Great Plains *annuus* and the cultivated *annuus* were all actively underway. Let us suppose that it was the white man who carried these sorts to California and stimulated *exilis* into producing *bolanderi* and started the remixing of *bolanderi* and *annuus.* In the Middle West he brought the prairies under extensive cultivation and allowed Great Plains *annuus* to spread much farther east and in greater quantity than it had done before. He brought it to Texas and began the mingling with the Texas species. He carried the single-headed sorts around the world, first as a curiosity and ornamental, finally as a great oil crop.

One little chapter in this story we do not have to guess; the facts have been recorded in detail. Mrs. T. D. A. Cockerell, the wife of a Professor at Boulder, Colorado, noticed one day a peculiar flower among the mongrel *annuus-petiolaris* weeds which are so ubiquitous in the cities and towns along the foot of the mountains. She collected the plant, and selected the best from among its seedlings; eventually she produced the red sunflower, now widespread as an ornamental. The dainty *cucumerifolius* was introduced into England from Texas as a plant for the flower border. Forms with paler yellow and deeper orange colors were selected and the charming little variety Stella came into being. These small-flowered sorts do well in the tropics and having been produced by an English seed house they have followed the Union Jack around the world. As a matter of fact, they persist even after the Union Jack has departed. Last winter I lectured at several universities in India and at every one of them I noted Stella blooming in bright profusion under the brilliant sun of an Indian winter. When I felt a little homesick I remembered that Stella eventually came from Texas and so I picked a bloom for my buttonhole and even used it sometimes to illustrate my remarks.

And now, just as this account goes to press, a possible new role for the sunflower is rumored from abroad. Two Swedish plant breeders have crossed the common sunflower with its perennial relative, the Jerusalem artichoke, and produced a hybrid with sweet sap in the stems. This gives promise of being not only good for forage, as they had hoped, but even of replacing sugar beets as a source of syrup and sugar.

This, in the barest outline, is the story of a comparatively simple crop and ornamental plant, the American sunflower. When we know the biological details with greater precision we can use them to illuminate the early prehistory of our own Southwest. We can trace the wanderings of people now unknown and produce decisive evidence for hotly disputed questions. When with something like this precision we are able to detail the development of a score or so of the world's major crops we shall be able to write the prehistory of man.

XII
Adventures in Chaos

NOW THAT WE HAVE COME to the last chapter it should be evident that this is largely a book about what we don't know. The one recurring theme which knits the various digressions together is that the ancient, everyday plants with which we spend our lives are virtually unknown; that for many of them we do not even have the simplest facts on record. Though a handful of us are beginning to look carefully into these matters, it is slow going. The study of the origin of cultivated plants is a field where there are few rules to begin with; one doesn't know ahead of time where even the partial answers lie; it is an adventure in apparent chaos.

There is only one branch of biology which is generally equipped to work in such a field. That is taxonomy. Though the taxonomists, as we have seen, have largely given up trying to do anything with these ancient plants, it is their general methods which must be applied if we are to work efficiently with such a complex and largely unexplored problem.

What is the taxonomic method? Well, it is the method of natural history. Marston Bates has recently written *The Nature of Natural History,* a whole book about what that consists of. It is easiest to define by example or by contrasting it with approaches more suitable for other types of problems. It customarily does not deal with exact units, it uses a minimum of mathematics, and until recently has had only a grudging respect from my fellow geneticists. Though I am not a taxonomist, though I am by nature fascinated with mathematics and have used mathematical methods in nearly all my scientific papers, I was fortunate fairly early in my scientific career to have acquired a real respect for the taxonomic method. It is probably no accident that Dr. Alfred C. Kinsey and I, who were graduate students together, should both be using the taxonomic method in places where it has not recently been generally used, he in the study of sex and I in the study of cultivated plants. Perhaps the best way to illustrate what I understand by the taxonomic method will be to tell something of the way I used it in beginning the study of maize.

When I began my work on corn I got in a few of our ordinary sorts and a few varieties from the Indians in the Southwest and grew them in my garden. In talking about them with my plant-breeder friends on the one hand and my taxonomic colleagues on the other, it gradually dawned on me over a period of several years that neither of these groups knew anything very much about the corn plant as such. The taxonomists were ignoring it because it was a cultivated plant; the plant breeders had been trained as cytologists or geneticists or statisticians; few of them had received even an elementary training in the natural history of the corn plant. They were making almost no attempt to see what they could find out about corn just by looking at corn plants and thinking about what they saw. Let me tell you for instance what one of our leading authorities was doing; he was a man for whose ability I have the greatest respect but he had no personal experience with the taxonomic method.

He was concerned with two related strains of corn, one with a higher average row number than the other. The plants also looked somewhat different and the leaves looked wider so he set out to get exact information on this point. He took the leaf above the ear, found its mid-point and measured the diameter there. He did this for fifty plants of each strain, calculated their averages, and determined mathematically the chances that such an average difference might have come about just anyhow and did not represent a really characteristic difference between the two sorts. The chances turned out to be pretty high and so in his published account of the matter he was careful to state that he had not certainly demonstrated a difference in the leaf, even after all this work.

Now the simple everyday facts about the leaves of the two strains were as follows. The two sets were of quite a different shape. One was broader and rounder at the base with a slight tendency to be wider in the middle and it was nearly always shorter than the other. One could easily demonstrate the differences in the two sorts, just by stripping off a series of leaves from one set and then a series from the other and laying them down on the earth of the cornfield in two parallel rows. With ten minutes of this kind of work it was clear that all the leaves of one strain were different from all the leaves of the other. Having seen half a dozen of each kind, one could classify at least nineteen out of twenty specimens which had been gathered by an assistant and brought in without a label. One of the results of this basic difference in shape was to make one set a little wider on the average, though the greater width did not necessarily show at the midpoint. With his uncertain reflection of the basic difference in shape, my friend had been struggling in his efforts to get a clear, mathematical answer. We may compare his problem rather exactly to that of separating a pile of oranges from a pile of apples. Apples and oranges being what they are, if we just measure their diameters it might take some pretty fancy mathematics to prove that the two sets of measurements probably came from different kinds of things. If we are willing to study the two fruits before we start out measuring them precisely, then we can see the myriad little differences between apples and oranges and any normal child can separate the two lots in a moment.

All of this was lost on my friend, not because he did not have a good mind, not because he was not industrious, but simply because he had not learned to look at a corn plant, or for that matter, any other plant. He was a most intelligent person, but he had never been given even the beginnings of a training in natural history. I found, to my horror, that although he had spent his adult life studying corn, he understood almost nothing of its technical architecture. Yet he was one of our very best corn geneticists and a man to whom students came, up until the day of his death, from all over the world. He had been convinced as a young man that the taxonomic method was old-fogyish, and he would have none of it.

A series of experiences of this sort convinced me that maize was an almost unknown plant. It was being completely ignored by the taxonomists who were technically equipped to study it; those who actually worked with it scorned such methods. I had not been technically trained in the taxonomic method, but several of my colleagues had, and most of my students. I set out in a groping fashion to find my way into the problem.

Now the reason why it is difficult to write about the taxonomic method is that in its broadest aspects it has never been described. Taxonomists are more like artists than like art critics; they practice their trade and don't discuss it. It was only by observing them at work and trying to translate what they were doing, to my own very different kind of a problem that I made very much progress. It appeared to me that one of their simple unwritten axioms seemed to be that different kinds of things vary in fundamentally different kinds of ways. The one thing which every taxonomist could tell you about maize was that it was a grass; to understand maize one would have to know something about grasses in general. I knew next to nothing about them technically, so I began in the simplest, most fundamental ways I could put my hands on. The English morphologist, Mrs. Agnes Arber, had just published a general volume on grasses. I read it from cover to cover. It seemed so simple that I wondered at her writing so elementary a book, but as the months went on I began to realize that like a good sonnet the book had more than just words — it transmitted an attitude, and with the new attitude one could look down whole new vistas of experience, and old facts took on new significance.

After these new ideas had seasoned for a few months I acquired another book, also deceptively simple, also by a woman, even by one of the same given name. It purported to be a field manual of grasses for the most elementary class of students. It would be better to describe Mrs. Agnes Chase's *First Book of Grasses* as being so fundamental in its simplicity that it is the scorn of college students but the delight of scholars. With this remarkable little book and a hand microscope I retired to the country for the summer and learned how to identify all the wild grasses of a small area. I found, to my joy, that though most botanists consider grasses difficult to understand they are really quite simple. The trouble springs merely from the fact that they are specialized. Their leaves look rather like the leaves of other plants and their roots are obviously roots but the rest of the plant, including the tassels, is so very peculiar that a familiarity with ordinary plants is not much help in understanding it. With Mrs. Chase to guide me these troubles disappeared and by the end of the first summer I could see that I was getting a real insight into these curiously specialized plants, many of which man has used for so many thousands of years.

With this kind of background I looked at the few collections of maize which had been sent me, and was amazed and bewildered by their variation. I grew two sets from Mexico, one from Jalisco, and one from Cherán, only fifty or sixty miles away in Michoacán. In each set there was more variation than in all the corn of Iowa, and yet as a whole each set was outside the range of variation of the other. Both proved to be different from other collections sent me from Guatemala. After several years of such preliminary fumblings I went to Mexico and lived in a suburb of Guadalajara where I could walk out into maize fields every day. By the end of the first week it was apparent that there was more variation in the corn of this one little town than in all of the maize in the United States. It varied from plant to plant, from variety to variety, and from field to field.

As the harvest season came on, I worked out methods of measuring a whole field. By semigraphical devices similar to those developed for measuring the "rubra" and "alba" of our previous hypothetical example, I succeeded at last in getting down on paper a record of one field which could be used to make precise comparisons with other fields, either in the same town or in other parts of Mexico.

Eventually the process of measuring maize fields worked itself out into a fairly standardized technique. I began by calling on my neighbors, explaining that I was studying corn and would like to measure a sample of their crop. As the season went on I spread out to neighboring villages and then traveled to other parts of Mexico. I often try to imagine what would happen to a visiting Mexican scholar, if alone and unassisted, speaking broken English, he traveled about the United States and stopped at small farms and asked to measure the corn in the crib! In Mexico I was always courteously received. I was usually given a chair and table to work at, even though it had to be brought from the neighbors'. Mexicans are by nature and tradition one of the world's most courteous and friendly people, but I suspect that part of my reception was due to the fact that I was studying maize. It is in Mexico not only the actual staff of life and the country's most important crop, it was anciently the center of their native religion. In most towns some of these ancient ceremonies have been taken over into the Catholic ritual and help to continue to dignify maize, and even mad foreigners who work with it in strange new ways.

Most of my work was done after the corn had been gathered and brought up to the house. Mexican houses open onto a central patio, or a series of interior patios in a big establishment. The corncrib usually opens upon one of these patios and I would sit there in the pleasantly warm winter sunshine, spreading the sample out on the tiled or earth pavement. I studied its variation, getting gradually a feeling for what each lot had in common in spite of the differences from ear to ear. The process was not so different from trying to characterize, shall we say, the people of Oshkosh versus the people of La Crosse, and then going on to differentiate both of these from the people in Albuquerque. I was indeed becoming the physical anthropologist of the maize plant!

The family usually stood around as I got out my photographic and measuring equipment, the children watching every move in open-eyed wonder. I learned to carry hard candies and sugar-coated gum in my pockets, and when I offered it to them they would edge forward shyly and take a piece. The family, particularly the old ladies, always brightened up at this attention to the young ones and the atmosphere immediately became even more co-operative. With metal calipers I measured, for each ear in the sample, the length and the width of the kernels, the width of the ear and of its basal shank, scored the relative hardness and pointedness of the kernels, and then recorded their colors. By trial, I determined that with twenty-five such ears, chosen at random from a corncrib, I could easily make a reasonably reliable record of the cornfield from which it came. Getting such a random sample was not always as easy as it sounds. Family pride often entered in and eager hands pawed through the pile in the crib and presented me with the very finest ears. I learned to accept these graciously, making the usual measurements and entering a checkmark on the data sheet, showing it was a selected ear and not one randomly chosen. One family was so helpful that I ended by measuring fifty ears, twenty-five selected by the family, and twenty-five at random. When measuring this double sample I thought I was exercising great forbearance; afterwards I came to realize that such dual records are priceless. They show both what the crop was, and what the family wanted it to be. By encouraging this kind of interference instead of just patiently putting up with it, I might have had an exact record of the Mexican fashion in corn, which could have been put to various useful purposes.

It takes a long time to measure twenty-five ears of corn. Eventually even a Mexican family would get bored at the sight of a strange foreigner measuring one ear after another, and they would go about their usual affairs. Since all the rooms opened on the patio, and since in the wintertime the patio was frequently the only part of the house which was comfortably warm, the daily life of the family flowed on for my education and amusement. Measuring and recording fifteen facts each for ear after ear is a terribly dull job, even for one really interested in all details of the corn plant. I began by feeling guilty about joining in everyday chitchat with the family. Towards the end of my stay in Mexico I began to suspect that this insight into the daily life of the common people was probably even more valuable than my precious statistics.

There was, for instance, the matter of the superstitious attitude towards variegated kernels. Very rarely in the United States but very commonly in Latin America, one finds ears of corn whose kernels are neither solid red nor solid white, but are covered with streaks, lines, and patches of alternate red and white. In Mexico these are usually known as *Sangre de Cristo,* literally "Blood of Christ." Now, though the common corn of western Mexico is white-kerneled, I noticed that in nearly every crib there would be a few ears of *Sangre de Cristo*. There was never a whole crib or field of this color, never more than a few ears, but these few turned up persistently in crib after crib. Talking about it with the old people in the patios, I learned that sometimes a few striped seeds were planted in each field as a sort of charm. Those who told me about the custom always claimed that while it was common practice in the community, their family, of course, never indulged in it; it was just an old-fashioned superstition.

Now that all this opportunity has slipped into the past, the one thing I regret is that I did not sit longer in these sunny patios and gossip more with the old folks. I would know more about maize if I had. It was such a pleasant way of passing the time that I felt kind of guilty about enjoying myself so completely, instead of getting more dull statistics. What I now realize I should have done was persistently to have led the conversation around to such matters as *Sangre de Cristo,* recording the substance of the conversations in a few notes at the edge of my data sheets, and writing it all down permanently when I got back to my lodgings in the evening. I was on the trail here of something of basic importance in understanding the origin and development of cultivated plants: the selection and perpetuation of a distinctive type because of its magic significance. I know enough about the matter now to be fairly certain that the inclusion of *Sangre de Cristo* kernels in fields of white corn must have been almost universal in western Mexico at the time I was there. By just enjoying myself a little more I might have had a real analysis of the custom, and not just vague suspicions. Oh, well, one learns!

I remember, now that it is forever too late, a good many little things of a similar sort which it would have been productive to have followed up. There was Señor Panteleone and his *maíz dulce*. He was the only man in our little town who continued to grow the old-fashioned native sweet corn, and I went to his garden in the corral behind his home several times to see it and talk to him about it. On each visit he kept telling me, "Now you realize, señor, that I grow this purely for a curiosity, purely for a curiosity." What was behind that reiterated explanation? Was he merely justifying to me in his own eyes the special trouble he was taking with an old-fashioned variety which his neighbors had abandoned, or was it something else? Was he merely putting his sophistication on record? I strongly suspect that it was something else, and something I would like to know about in detail. In South America these ancient sweet varieties are grown for making specially potent *chichas* and other alcoholic and semi-alcoholic beverages. Remembering Señor Panteleone and his little house well off at the edge of our town, I think it was quite likely that he was going to all of this trouble of having a patch of *maíz dulce* in order to indulge himself with a little illegal corn beer made in some ancient way. Had I been more receptive I could have gained his confidence and learned much from him.

One unexpected fact became increasingly apparent as I studied more and more Mexican maize. If I traveled westward from my home in San Pedro Tlaquepaque, the corn all had narrow ears, it was borne on slender plants with tough narrow leaves, and the tassel branches were long and slender. To the east, the corn type changed rapidly from field to field and from village to village. Within a hundred miles all the corn, though still variable, was entirely different from the maize about Tlaquepaque. Not only was it different on the average; any one ear in these villages to the eastward would be quite unlike any of the ears in the western villages. It was no longer slender and tough. The leaves were wide and broke easily in the wind; frequently they were hairy and rough. The tassels were shorter and stubbier, the ears were shorter and thicker, the kernels smaller and more pointed. Maize in Mexico seemed, like mankind, to be made up of vaguely defined races which occupied different areas and graded imperceptibly into one another. Sometimes the transition could be very sharp as when Oscar Lewis helped me sample one hundred fields in the township of Tepoztlán and we found only a few fields which were intermediate, all the others being clearly of one race or the other. From archaeological evidence and from early Spanish accounts we learned that this division into different races was of long standing, going back at least five hundred years and probably very much longer. Among the most primitive varieties in each race were the popcorns. In western Mexico the common village popcorn was called *maíz reventador* (that is to say, "the corn which explodes"). It exhibited all the characteristics we had come to associate with the western race, some of them in an exaggerated form. The ears were slender, almost the shape of a cigar, and not very much larger than one, their tiny white kernels set smoothly together like tiles in a pavement. When I went to central Mexico to study the corn there, I carried one or two ears of *maíz reventador* in my brief case and showed them to farmers and maize experts. None had ever seen anything like them. Their own popcorns had short stubby ears, like little pine cones, the kernels so sharply pointed that one of the commonest local names for the variety was "rice corn." When I carried some of these *maíz de arroz* ears back home to Tlaquepaque they created equal amazement among my farmer neighbors.

Now the main point to be made here is not that the maize of Mexico is divided into a number of races and subraces, much like mankind. If that interests you, get the handsome report of the Mexican Department of Agriculture which, with the co-operation of the Rockefeller Foundation, has gone on to describe their importance to ethnological problems and to modern maize breeding. No, the reason for going into the matter here is simply that it is a further demonstration of the kind of discovery made by the use of the taxonomic method. Before I began my work little or nothing had ever been said about races of maize. Corn was supposed to be sharply and uniformly divided into flints, sweets, pops, and the like, with little or no geographical differentiation. I had no notion when I began the work that I was going to run into any such phenomenon. It is not the sort of thing which is apparent in the specialized agriculture of the United States and of northern Europe, and was outside not only my own experience but that of other American and European scientists who had been studying our crop plants. The division of *Zea mays* into a number of races and subraces is the most fundamental fact about corn, but had I used some more new-fangled method of studying it, I might have gone on for years before the division into races forced itself upon my attention. As it was, the use of the taxonomic method let the facts almost speak for themselves, and it quickly emerged from the study.

Others came with time. One conclusion was particularly interesting because it was directly opposed to what I had been led to believe. It concerned the purity of Indian varieties of corn, which are ordinarily described as very mixed compared to modern ones. In Mexico I worked almost exclusively with farmers of European or partly European ancestry. Even those who had strikingly Indian features were mostly Spanish-speaking and did not consider themselves Indians. In Guatemala I worked with such people but also with Indians who had retained their old languages and their own cultures. I found, to my surprise, that their cornfields had been more rigidly selected for type than those of their Latin-speaking neighbors. Their fields were quite as true to type as had been prize-winning American cornfields in the great corn-show era when the American farmer was paying exquisite attention to such fancy show points as uniformity. This fact was amazing, considering the great variability of Guatemalan maize as a whole, and the fact that corn crosses so easily. A little pollen blown from one field to another will introduce mongrel germ plasm. Only the most finicky selection of seed ears and the pulling out of plants which are off type could keep a variety pure under such conditions. Yet for Mexico and Guatemala and our own Southwest the evidence is clear: wherever the old Indian cultures have survived most completely the corn is least variable within the variety.

Much later I grew a collection of corn made among an even more primitive people, the Naga of Assam, whom some ethnologists describe as still living in the Stone Age in so far as their daily life is concerned. Each tribe had several different varieties which were sharply different from one another, yet within the variety there were almost no differences from plant to plant. Furthermore, some of the most distinctive of these varieties were grown not only by different families but by different tribes, in different areas. Only a fanatical adherence to an ideal type could have kept these varieties so pure when they were being traded from family to family and from tribe to tribe. It is apparently not true, as has so frequently been stated, that the most primitive people have the most variable varieties. Quite the opposite. It is rather those natives most frequently seen by travelers, the ones who live along modern highways and near big cities, the ones whose ancient cultures have most completely broken down, who have given rise to the impression that primitive people are careless plant breeders.

Well, perhaps this story of the way I began to study maize and its varied uses will give you some idea of what I mean by the taxonomic method and why I have described it as an adventure in chaos. I could not, you see, have known when I started in that maize would be so variable that I would have to work out methods of measuring whole fields. Even after these methods were well developed I did not realize that I would eventually use them to ascertain the degree to which different peoples kept their maize varieties true to type. It never occurred to me until the work was well under way that the attitude of the growers towards what they wanted the crop to be like was one of the points on which I should be getting precise information. In using the taxonomic method one begins by making a broad survey of the whole problem in its widest aspects, and then gradually as he becomes increasingly confident about where the most significant data are to be found, he narrows down his investigations to answer fairly precise questions. It strikes me as a most valuable kind of training for those who live in these modern times. The world in which we now find ourselves seems pretty chaotic. As in a taxonomic problem, one has to look here and there for little vestiges of apparent order and work out experimentally from them, with such techniques as can be devised as one goes along. He gets on best who has a flexible mind, who does not think in hard and fast terms, who is used to searching through apparently chaotic material for such beginnings of understanding as can be built up in a tentative way.

During the first five years that I studied maize and its wild relatives I continued to struggle with the problem of making an adequate record of the few samples I grew every year. When the plants were alive and in good health it was fairly simple to study them. One could walk back and forth repeatedly through the experimental plot and compare the various sorts feature by feature. It was not so easy to make comparisons with plants grown last year or the year before that. To be sure, one could make measurements and photographs or make herbarium specimens, but none of these methods worked out very well in practice. It was impossible to anticipate which measurements would be most advantageous, the photographs never seemed to show the details one would most like to use for comparison, and the herbarium specimens, even of just the tassel alone, were distressingly bulky. Making an adequate herbarium specimen of a corn plant, I remarked petulantly, was like trying to stable a camel in a dog kennel.

With a little ingenuity it was possible to fuse all these techniques and produce a permanent and usable record. Careful photographs were taken, to scale, against a white background. As finally worked out, this was a typical bit of scientific technique, merely a fussy and precise adaptation of something which had previously been done without very much precision. A large white wall about the size of the end of a garage was erected in the shade near the experimental plot. Black lines were painted across it at regular intervals, and the year and the location were printed in one corner in large letters. Corn plants were dug up from the experimental plot and held tight against the wall by a line of deheaded shinglenails. The name and number of the variety were printed on a white card and thumbtacked beside the specimen. Then the whole thing was photographed. Against the white background, with no confusing shadows, the entire ground plan of the corn plant showed up clearly. The exact number of leaves, their size and proportion, the number of ears, their height above the ground, the size and shape of the tassel—all these and many other technical details could be determined exactly in each photograph. The labels had been photographed right on the film and there was little possibility of confusing the record with one made of another plant or at another place. A careful print of this photograph was the central feature of the amplified herbarium record. It was glued to a standard-size herbarium sheet, along with an enlarged photograph of the tassel, made in the same fashion, a diagram showing the number of internodes on the stem and their relative lengths, notes on the color of the stamens and the tassels, herbarium specimens of the central spike and the lowermost branch of the tassel, and a summary of information about the chromosomes derived from studying microscopic preparations (when they were available). Photographs of the ears and the kernels were added to the sheet, and when as occasionally happened, the seed was from a field which I had measured, a diagram of that field went into the record. Aside from the fact that it was on an herbarium sheet and could be filed in a standard herbarium case, this concise record of a corn plant was in many ways much more like a notebook than an herbarium specimen. When I learned what an effective technique it was, I started demonstrating it to various scientific friends. The taxonomists, to my chagrin, showed no interest whatever; some of them even seemed positively annoyed that one of their folkways had been so debased. The corn-breeders, on the other hand, were quite tolerant about the idea and one corn-breeding company even installed a couple of herbarium cases and made specimens of its own, and exchanged some of these for some of mine. Very slowly the idea began to spread. Gradually I tried it out with other kinds of crop plants and found that it could be adapted to almost anything. Though this seemed like a technical advance too tiny for scientific notice, it became increasingly clear that these amplified herbarium specimens could be put to various useful purposes, and would yield exact and decisive information on many kinds of problems. So, after five years' trial I finally gave the device a formal description and named it "the inclusive herbarium."

Now, as all the maize work developed I became increasingly aware of the basic problems we have been discussing in this book, the lack of precise information about our commonest plants, the pressing need of extending these studies of maize to other crops and to ornamental plants and weeds. I succeeded in interesting one of my students in cucurbits, and one of my junior colleagues in sunflowers. Jonathan Sauer arrived to study with me, and insisted on tackling one of the hardest problems of all, the grain amaranths. With the curator of our grass herbarium I started a comprehensive collection of the various sorts of Sorghum. All these efforts, however, barely touched the main problem. The proper kind of specimens with which to carry on such studies simply did not exist. If we were to get very far with our understanding of cultivated plants and weeds we would have to start at the bottom and build up "inclusive herbaria" devoted to their study. We couldn't expect such institutions just to start overnight and get running full tilt, but somewhere, it seemed to me, a small beginning should be made at the systematic collection of these all-important plants.

The Quakers have an interesting theory about such urgent and purposeful desires, and being a Quaker, I am well acquainted with it. The really essential thing, according to the old-fashioned Quaker phraseology, is "to have a concern." If one is really concerned, then "a way will be opened." I suppose there are important exceptions to this widely held Quaker belief, but in my own experience, it is amazing the number of times it has become literally true, including as we shall see, this concern for an inclusive herbarium of cultivated plants. To make an effective beginning would not take very much money or effort, but even that little seemed beyond any possible resources I could lay my hands on. One needed a small experimental plot somewhere in the tropics — in the tropics, because that is where agriculture apparently got started and our oldest civilizations and oldest crops are to be found in the tropics or subtropics. Aside from the plot, one needed only facilities for making and storing herbarium specimens and doing photographic work, and for healthy living in a tropical or subtropical environment. A good library and colleagues acquainted with the tropics, though not absolutely necessary, would be helpful in furthering the project. All of this added up to more money than I could hope to obtain from any one source, but I nursed my concern along until finally it dawned on me that by enlisting the help of the Escuela Agrícola Panamericana I could probably begin the actual assembly of an inclusive herbarium.

The Escuela Agrícola is a remarkable institution. It was set up somewhat over a decade ago by the United Fruit Company as one means of bringing about a more rational development of natural resources in Central America. Sixty to seventy boys, mostly the sons of small farmers, are admitted each year and are given a practical three-year course in tropical agriculture. The boys come from a dozen different republics, the staff is partly from the United States and partly from various Latin American countries, and the whole establishment is located in a fertile valley in central Honduras, in one of the healthiest spots in that part of the world. It has a good scientific library, an excellent herbarium, and ample experimental fields.

Turning over in my mind this urgent concern for an inclusive herbarium, I suddenly realized that with a small grant of three or four thousand dollars, I could easily get the project started provided the Escuela Agrícola would co-operate, and I was fairly certain they might. I knew the director and some of the staff and it struck me that having such a project carried on there might be stimulating and advantageous to their own concerns as well as to mine. Fortunately they agreed and with a grant from the Guggenheim Foundation we got under way. It has now been going over a year and both the school and I are encouraged by what has already been accomplished.

The first thing was to select one of my students who had some training in taxonomy but had much wider interests and did not wish to settle down to the sedate existence of an ordinary herbarium curator. The school turned over a plot of ground about the size of a small suburban garden and we cleared it of rocks and it was fenced in. We began by planting a few collections of things which were already at hand from various parts of Latin America; and when the United Nations sent me to India for a plant-breeding conference, I picked up a few things there. The idea was not to make as many amplified herbarium specimens as possible, but to feel our way along and work out the most effective way of operating such a project. My student, fortunately, was already very interested in finding out something about the common bean, so he grew a number of collections of beans. Within the first year this developed into a bean-breeding project for the school, and he now spends half his time working for the school as a bean breeder and half working on our inclusive herbarium.

We plan to make somewhere between twenty-five and fifty specimens a year, each one in sets of three. The first set will stay at school, the second will come to me, and the third will be exchanged with some other student of cultivated plants. Our main job, we feel, is not making the specimens, but studying the plants thoroughly so that by the time they are mature we have found out in what ways they vary from plant to plant, and how best to record these facts. The school has built us a big white photographic background not far away from our experimental plot, on the wall of the school's sugar mill. One of the crops we concentrated on the first year was grain amaranths, since Jonathan Sauer had been monographing them and had collected several lots of seed from native sources. They were grown in triplicate in various parts of the garden to see how much differences in soil and site affected their development. We found that aside from the height of the plants there were few differences of this sort but that the various collections differed a great deal in their plant-to-plant variability. One or two cultures were so diverse that we made detailed plant-to-plant records. For the others we dug up half a dozen mature plants, set them up against our prepared background and photographed the whole lot at once. With this photograph, a carefully pressed leaf and inflorescence, and a few notes about the variation in color, we shall have the world's only effective specimens of grain amaranths. Sauer's notes, down in one corner will tell us where the crop was being grown and for what purpose. Our notes, photographs, and specimens, will demonstrate what the plants were like and how much variation there was in the lot. One such sheet will be worth hundreds of ordinary specimens to any serious student of this ancient crop.

Not all our time and thought is centered around the experimental plot. It is our business to learn how to make an effective record of the plants in man's transported landscapes, and we will probably be most efficient by not sticking too closely to some hard and fast formula in starting out on this quest. On my last visit, we set out to study the native gardens at the homes of small farmers in the upper end of the valley. These are little dooryard plots something like the mixtures of orchard-vegetable-flower garden I have previously described for Santa Lucia in Guatemala. The very first one we studied extended the world's understanding of the grain amaranths. Though these plants had not previously been reported from Honduras, we found this little garden was peppered with the attractive bright-red plumes of an amaranth. Inquiry revealed that the family were not growing it for seed, and did not even know that it could be used for nourishment. To them it was a magic plant, whose seeds, wrapped up in little packets and carried close to the chest, had the power to ward off a cold or to cure one after it had started. During the second year of our project, we are planning to make friends with several of these cottagers and make as complete a record as possible of all their dooryard plants, ornamentals, vegetables, fruits, and weeds. It will be an accurate record not only of the plant itself, but of its relation to the family, why it is there, what they do with it, and how it fits into their lives.

Another of our attempts to work toward an inclusive herbarium is concerned with such gigantic plants as palms and bananas, which have proved difficult to file in any ordinary herbarium. L. H. Bailey, the veteran American student of cultivated plants, has developed methods of pressing these gigantic leaves and inflorescences so that they can be stored in regular herbarium cases, but his specimens take up so much room that no ordinary institution is going to be able to preserve many specimens. It seemed to me that the methods I had worked out for corn should be extended to palms, since no consideration of the cultivated plants of the world is going to be complete until the palms are included. For man in the tropics they are as important as all of the rest of the vegetation put together. They supply food, forage, clothing, building materials, thatch. They are among man's oldest associates. Agriculture apparently started in the tropics and some of the palms must be among our oldest domesticates. They are becoming increasingly important in modern technology, as should any plant which is capable of bearing several tons of fruit a year, as do some of the palms. For oils, fats, waxes, as well as for human food, they are being imported into the Temperate Zone in increasing abundance. Some of these technological users have already run into the hard fact that the palms are extremely variable and before they can be utilized effectively they will have to be much better understood than they are at present. We thought it was high time to make another attempt to stable camels in dog kennels, and see if we couldn't work out an effective way of recording the technical facts about one palm tree on just one small herbarium sheet.

The campus and pastures of the Escuela Agrícola were studded with coyol palms, which seemed an excellent kind to begin our study with. There were several hundred specimens, each one about the height of a telephone pole, and they obviously varied a good deal from tree to tree in the development of heavy spines on the trunk and leaves, in the way the leaf bases clung to the trunk after the leaves had fallen, and in the delicacy of the ultimate segments into which the huge leaves are divided. Though you and I do not appreciate the fruits, they are relished by cattle and are even munched enthusiastically by children of the poorer classes. To me they taste much as they look, like a smallish green golf ball with a flesh only slightly more mucilaginous than the commercial sort and very faintly sweet to the taste. Coyol groves such as that at the Escuela Agrícola usually represent either deliberately established orchards or are the incidental outcome of ancient villages.

The coyol, in other words, is one of those plants which have been carried around by man and have in the process become so variable, their natural division into species and subspecies so blurred, that their classification presents a special problem. Specimens from two different palm trees on the grounds of the Escuela were sent to one of the authorities on palms and came back with the note that in his opinion they represented two species in different sections of the genus! To us, that suggested that if the coyol couldn't be dealt with satisfactorily by old-fashioned methods then we might be able to make sense out of the problem by diagramming the variation of the whole orchard, using methods similar to those I have described in Chapter 6. First we would have to learn how to get the essential facts about one palm down on one herbarium sheet; when that was finished we could go on and make a diagram of variation in the whole orchard which would go on a second sheet.

We began by spending one or two evenings walking from tree to tree noticing the variability and making careful comparisons, character by character. After a week of such occasional skirmishes we hunted up a grub hoe and a couple of machetes and proceeded to fell one of the old specimens which the dairy department had asked to have removed from the calf yard. The process served the double purpose of getting the palm tree down with such equipment as was readily available, and teaching us a good deal more about the structure of palms. Some palms are said to have a trunk which is easily severed with a machete. Not the coyol. When we tried a few preliminary whacks it gave off a metallic sound and the machetes bounced back in our faces. It was more vulnerable at the roots. There are hundreds of these wiry little cables about the width of your finger. They radiate out from the base of the trunk in all directions, so with grub hoe and machetes we dug a two-foot trench immediately around the tree, cutting all the superficial roots which radiate horizontally and even getting down to some of those which go at more of an angle. Then after dinner we rounded up most of the junior staff and fixed two ropes around the palm trunk as high as we could reach. With two men pulling on the end of each rope and everyone else pushing against the trunk, the big tree was down in a few seconds and we spent the rest of the evening dissecting it with our machetes.

The big leaves, each one about the length and width of a narrow hallway, were carried to our prepared photographic background to be photographed to scale. Thin sections were cut out of the midrib of the leaf, each one an exact record of the way that organ was channeled and the extent to which it was provided with spines and hairs on the upper and lower sides. We cut out and photographed the big inflorescences and the woodeny brown spathes around them, the latter almost the size and shape of old-fashioned bread-kneading trays, though as thin as cardboard. When the photographs had been taken we reserved a few of the thin branches in the inflorescence, some of the ultimate leaflets or pinnae, and characteristic spines from the trunk, and pressed them carefully. Measurements were made of the trunk and of the leaves and leaflets. Finally all these data were brought together and assembled on standard herbarium sheets. Each sheet had photographs of the tree, the leaves, the inflorescence with its fruits, and the spathe. It had a little piece of the spathe, slightly larger than a commemorative postage stamp, cut out of the palm near its apex. It had one or two leaflets, a trunk spine, and sections of the midrib at base and center. Measurements of trunk, leaves, and leaflets were at one edge of the sheet.

For all practical purposes we had succeeded in stabling our camel in a dog kennel. The essential facts with regard to this one palm tree had all been brought together on one sheet of herbarium paper. Experience would teach us how to make an even better record, but this at least was a beginning. In time we would find ways and means of sampling the world's most important palms. We had had a concern, and a way was being opened.

This detailed story of how we cut down a palm tree and measured some of the pieces and photographed the rest does not sound like modern science as one reads about it in the Sunday newspapers. It is all very simple, the equipment (aside from the camera and light meter) is all very old-fashioned. There are no vacuum tubes, no cloud chambers, no microscopes, no elaborate apparatus for keeping everything at a constant temperature. Well, one of the difficulties of finding support for science in a modern democracy is that only certain phases of science have an immediate popular appeal. Frequently a really important scientific advance is nothing more exciting than getting precise data on an old problem, newly seen to be peculiarly significant.

So it is with the palms and the other plants which have traveled about with us for so long. It is now apparent that really precise data about them can be the key to many problems. These plants that have been with us since the Stone Age, that may have even come into being under our influence, that have shaped our own destinies, will give us the data from which we can write the story of man. Learning how to get one palm tree down on paper with new precision is just one step. Having taken that, we may hope to go on and learn to make effective records of whole orchards of coyol palms. With two or three such collections we may then extrapolate precisely from the coyol of today to what the coyol must have been like before man appeared. We may even hope to determine its progenitors and ascertain where it first associated with man and by what routes it was carried about. From one kind of palm we may go on to others and to a larger understanding of man's history and destiny in the tropics.

The story of our first coyol and how we at length stabled it efficiently on herbarium sheets is significant because to us at least, it marks the turning point at which order first became definitely apparent in the chaos. From about that moment we could see our way ahead with some confidence. The years of adventurous blundering were largely behind us and an era of orderly advance lay just ahead.

Epilogue

HAVING A BOOK PUBLISHED can change one's life in as many ways as having a baby in the family. Shortly after *Plants, Man and Life* appeared, it was reviewed in the journal, *Landscape*. This led me to J. B. Jackson, the able editor of that remarkable publication. With a piquant combination of sharp criticism and flattering appreciation, he charmed out of me a series of book reviews and short essays. Most of these are related in one way or another to matters discussed in this book and references to them are included under Suggested Readings. One of them takes up in more detail the "dumpheap theory" of the origin of cultivated plants. Several touch in one way or another on the acceptance of cities as places to live right in the middle of; an attitude that is part of Mexico's Spanish heritage.

Three of the essays describe how I learned enough from my Mexican neighbors to have lived serenely years later in a big, moderately priced St. Louis apartment hotel for six months, as a naturalist happy in learning from our apartment windows: new things about bird life in the big city, the serpentine course (unparalleled in its loopings) of the Missouri Pacific railroad as it starts southward, the progress of small thunderheads across the city on a day of little thunder showers, the dynamics of winter sunsets when the sun is so low in the sky that the observer needs to be well over a hundred feet above the ground level to learn very much, and the heights and extents of autumnal morning fogs. I was again spurred on to study and to teach the natural history of cities along with that of seashores and spring woodlands. Some of those who liked this book have found deeper satisfactions in these *Landscape* papers. The editorial address is Box 2323, Santa Fe, New Mexico 87501.

The thinking behind the pictorial diagrams of figures 8 through 10 was given its most thorough-going discussion in the *American Journal of Botany,* 43:882–889, December 1956. For this issue I wrote an essay, "Natural History, Statistics, and Applied Mathematics," that pursued further some of the ideas expressed in *Plants, Man and Life*. Currently, it is of interest to a few of those who are adapting computer techniques for what this essay describes as "pattern data." Though parts of it are technical, the long introductory portion will be understood by most readers.

Suggested Readings

Ames, Oakes: *Economic Annuals and Human Cultures.* 153 pp. Botanical Museum of Harvard University, Cambridge, Mass., 1939.

Discussed in detail in Chapter VIII.

Anderson, Edgar: "Natural History, Statistics, and Applied Mathematics," *Am. Jour. Bot.,* 43:882–889 (Dec. 1956).

Discussed in Epilogue.

Baker, Herbert G.: "The Evolution of the Cultivated Kapok Tree, a Probable West African Product," in *Ecology and Economic Development in Tropical Africa.* Pp. 185–216. University of California, Research Series 9, 1965.

An outstanding example of how an investigation of a minor cultivated plant can illuminate prehistory.

Burkill, I. H., *et al.: A Dictionary of the Economic Products of the Malay Peninsula.* Oxford University Press, London, 1935.

Burkill's two closely packed volumes might well be described as "Watt up to date." In addition to the information available in Watt, Burkill's work contains more distilled from his own experience, and a summary of the information which can be sifted from the world's technical literature.

———: "Habits of Man and the Origins of the Cultivated Plants of the Old World," *Proceedings of the Linnean Society of London,* 164:12–42 (1952).

An extension of Burkill's Hooker Lecture of November 22, 1952. Burkill is the greatest ethnobotanist the world has seen. He had de Candolle to build on, and he had been director of the fabulous botanical garden at Singapore. He also had an integrated understanding of his subject in the garden, the market place, the back country, and in the world's scientific literature.

GORDON, B. LE ROY: *Human Geography and Ecology in the Sinú Country of Colombia.* 136 pp. University of California Publications: Ibero-Americana, Vol. 39. University of California Press, Berkeley and Los Angeles, 1957.

The link between the prehistory of the Caribbean and of the Pacific.

HARLAN, HARRY V.: *One Man's Life with Barley.* 223 pp. Exposition Press, N. Y., 1957.

An authoritative (but off the cuff) account of the history and classification of the barleys of the world by the foremost authority. An amazing potpourri of his adventures in the back corners of the world, the prehistoric role of barley and its classification, breeding, and study, shrewd observations on the folkways of fellow scientists and on life in general and in government circles. With a moving (and lively) eight-page introduction by his long-time assistant, Mary Martini.

HARRIS, DAVID R.: *Plants, Animals, and Man in the Outer Leeward Islands, West Indies: An Ecological Study of Antigua, Barbuda, and Anguilla.* 164 pp. University of California Publications in Geography, Vol. 18. University of California Press, Berkeley and Los Angeles, 1965.

An inquiry into domestication on the Atlantic fringe of the Caribbean.

HEISER, CHARLES B., JR.: "Natural Hybridization with Particular Reference to Introgression," *Bot. Rev.,* 15:645–687 (Dec. 1949).

As this book goes to press, Professor Heiser is actively at work on a monograph of the sunflowers (Helianthus) wild, weed, and cultivated, annual and perennial.

HUTCHINSON, SIR JOSEPH: "The History and Relationships of the World's Cottons," *Endeavor,* 21:5–15 (1962).

After a lifetime of intensive investigation around the world by Sir Joseph and his associates, the story of cotton has grown out of dogma into an orderly synthesis of historical, anthropological, and botanical study. The main outlines of the interrelationships between the world's wild, weed, and cultivated cottons are now generally agreed upon, though certain points are still matters for debate. Several major points now universally agreed to were thought too bizarre for serious consideration when these intensive studies were first initiated.

PEATTIE, DONALD C.: *Cargoes and Harvests.* 311 pp. Appleton & Co., N.Y., 1926.

Like my own book, an outgrowth of Oakes Ames' course in economic botany.

SAFFORD, WILLIAM E.: *Useful Plants of Guam.* 416 pp. Contributions from U.S. National Herbarium, Vol. IX. Government Printing Office, Washington, 1905.

Under this modest title is hidden one of the world's most fascinating volumes. The author, who apparently came as close to knowing everything about everything as is possible in modern times, was professionally both a botanist in the United States Department of Agriculture and a lieutenant in the United States Navy. In this latter capacity he served for a year as assistant governor of Guam. In somewhat over four hundred pages he not only takes up all the native and crop plants of any importance, but also touches on such subjects as the history of pirates in the Pacific, how floating seeds led to the discovery of ocean currents, the grammar of the native language, the actual anatomical means by which stinging plants attain their devilish ends, and the aspect of the various kinds of tropical vegetation on the island, each of these digressions being developed with finicky regard for accuracy and appropriately embellished with authoritative footnotes.

Something of its contents and style are indicated by the opening words of the introduction:

"During a series of cruises in the Pacific Ocean the routine of my official duties was pleasantly broken by frequent excursions on shore for the purpose of collecting material for the United States National Museum, as well as for recreation. While sitting in native huts and while wading upon coral reefs, traversing forests and climbing mountains, I interested myself in taking notes on the languages and customs of the natives, their arts, medicines, food materials, and the manner of preparing them, and the origin of their dyes, paints, fibers for fishing nets and lines, materials for mat making and thatching, woods used in constructing their houses and canoes, and gums and resins used in calking.... It occurred to me, therefore, that a popular work on the useful plants of Polynesia would be welcome, and I set out accordingly to gather together such information as I could for this purpose."

SALAMAN, REDCLIFFE N.: *The History and Social Influence of the Potato,* with a chapter on industrial uses by W. G. Burton. 685 pp. Cambridge University Press, London, 1949.

This fascinating volume is long on what happened to the potato after it got to Europe and short on how the potato ever got started as a primitive American crop. It is clearly written and well indexed, and provided with interesting and pertinent illustrations.

SAUER, CARL O.: *Agricultural Origins and Dispersals.* 110 pp. The American Geographical Society, N. Y., 1952. Given as one series of the Bowman Memorial Lectures in New York City in 1952.

A senior scholar shows his expert musings about the very beginnings of agriculture.

———: *The Early Spanish Main.* 306 pp. University of California Press, Berkeley and Los Angeles, 1966.

The same scholar, years later, presents a meticulous volume on the learned world's first contacts with the civilizations (and hence the agricultures) of the New World. With the persistence of genius he did not put his account together until (1) he had visited all significant sites in the Old World and the New, making his own penetrating observations of their geomorphology from the air and on the ground, and (2) he had studied the original accounts and had unearthed unnoticed (or forgotten) ones in European archives.

The anguish of book reviewers who were having to realign their geographical and historical backgrounds has obscured the contributions this volume brings to ethnography, ethnology, anthropology, the ecology of the tropics, the origin and development of agriculture, and the Melanesian threads in the tangled skein of early voyages about the Pacific.

SAUER, JONATHAN D.: "The Grain Amaranths: A Survey of Their History and Classification," *Annals of the Missouri Botanical Garden,* 37:561–632 (Nov. 1950).

Discussed on pages 159–160, above.

SCHERY, ROBERT W.: *Plants for Man.* 564 pp. Prentice-Hall, N.Y., 1952.

Comprehensive, accurate, and superbly illustrated.

SCHIEMANN, E.: *Enstehung der Kulturpflanzen.* 377 pp. Handbuch der Vererbungswissenschaft (E. Baur and M. Hartmann, eds.), Vol. III, Lieferung 15. Gebrüder Borntraeger, Berlin, 1932.

Though this is in German and is solid reading (even for a German monograph), it is included because it is as yet the only summary we have of a most important field of work. It brings together modern studies of the cytology and genetics of cultivated plants and the problem of their origin and development. Unlike most cytologists and geneticists Miss Schiemann was well prepared to perform such a technical synthesis. Born in a family which knew well several distinguished anthropologists and explorers, most of her professional life has been spent in laboratories of genetics and she herself became an authority on the relationships of primitive and modern wheats and the history of the strawberry.

VAVILOV, N. I.: *The Origin, Variation, Immunity, and Breeding of Cultivated Plants.* Trans. from the Russian by K. Starr Chester. 364 pp. Chronica Botanica Co., Waltham, Mass., 1951.

Vavilov's original papers are widely scattered and though many of them are in English or have English summaries, they are mostly in publications not to be found in the average public library. In this generally available volume the best of Vavilov's writings have been ingeniously and tastefully brought together by the distinguished editor of *Chronica Botanica.* It includes Vavilov's major contributions to biological thought along with summaries of some of the actual data on which he based his far-reaching conclusions.

WALKER, HELEN M.: *Elementary Statistical Methods.* Rev. ed., 302 pp. Holt, Rinehart & Winston, N. Y., 1958.

WATT, SIR GEORGE: *A Dictionary of the Economic Products of India.* (The "big Watt.") Calcutta, 1885–1893.

———: *The Commercial Products of India.* (The "little Watt.") John Murray, London, 1908.

While neither the "big Watt" nor the "little Watt" is the kind of book one can just walk into a bookstore and buy, the "little Watt" is generally available in technological libraries and the "big Watt" can be consulted in any first-class botanical library. Of all the works cited in this reading list the "big Watt," though published in 1885, is closest to the subject matter of *Plants, Man and Life.* Sir George Watt, a professor in the Bengal Educational Department serving on special duty with the Department of Revenue and Agriculture, in India, instituted the wide survey of plant, animal, and mineral products on which his two works are based. The "big Watt" contains the actual reports made according to his specifications as they came in to him from all over India. Though many of the reports overlap, they were apparently printed in full so that after reading through the material on any one crop plant one gets the feeling of having interviewed a succession of specialists. Though many of the more technical portions are put in fine print, the collection fills six volumes with sheets of almost folio size. However, it tells one everything that a driving administrator could bring together: the native names, the native uses, the yields, the history, the classification—not only for India's major crops, but for most of the minor ones as well. Were I to mark with red the days in my professional life when really important discoveries were made, one of the first candidates would be the afternoon when in answering a request for information I blundered about in the library and found for myself this incredible storehouse of facts.

WILLIS, J. C.: *A Dictionary of the Flowering Plants and Ferns.* 6th ed., rev., 752 + liv pp. Cambridge University Press, London, 1931.

Willis appears on this list, not because his little handbook is anything to sit down with and read but because for a small and widely available book it is more inclusive than any other. If, for instance, someone tells you that a fruit in a native market is a monkey apple, you can refer to Willis and learn that in the West Indies this name is applied to a species of Anona. By turning to the entry under Anona one learns that four other species of Anona from the American tropics are widely grown as tropical fruits (the cherimoya, the sweet sop, the sour sop, the bullock's-heart) but that the monkey apple is not important enough to rate any further mention.

WISER, CHARLOTTE V.: "The Food of a Hindu Village," *Annals of the Missouri Botanical Garden,* 42:303–412 (Dec. 1955).

An ethnobotanical classic. An accurate, detailed, and perceptive analysis of what plants were grown for what purposes. Since India is a land of villages (with a few enormous cities), this is background for understanding the India that few intellectuals know much about.

Glossary

Agronomy The scientific study of the history, breeding, and management of crop plants.

Artifact Some object produced through human contriving. The term is often used in contradistinction to objects produced *without* human intervention. See Weeds as Artifacts, p. 9.

Barrow Common term in England for an ancient burial mound.

Bell beaker One type of ceremonial drinking cup.

Butt The base of an ear of Indian corn (maize) where it is attached to the parent plant.

Corn The common grain of a country. The English "ear of corn" and the U.S. "head of wheat" are two different names for the same object. In the U.S., Indian corn (technically maize) has lost the identifying prefix and is called corn in common speech.

Cucurbit Any member of the gourd family (Cucurbitaceae): gourds, pumpkins, squashes, cucumbers, and the like.

Diploid Two-fold, see p. 54.

Domesticate This verb has, in laboratory slang, become a noun designating a domesticated animal or purposefully cultivated plant.

Endemic A species or variety of plant or animal that has long been known from a particular region. By extension this term is used to refer to species which are found only in a relatively small region.

Fallow A field tilled and weeded but uncropped to increase fertility and discourage weeds.

Glume One of the chief parts of the flower in certain kinds of plants, often comparatively conspicuous in grasses (p. 15).

Glyph A standardized pictorial symbol (as in the ancient Mexican picture writing).

Homologous Having the same relation; see discussion of "homologous variation" with detailed examples, pp. 75–78.

Maize The precise common name for Indian corn, botanically classified as *Zea mays.*

Midden A prehistoric refuse and rubbish heap.

Millet Any one of several kinds of grain crops which as human food are less palatable than rice, wheat, or barley. They belong to different genera and are no more closely related than are apples, strawberries, and plums, all of which belong in different parts of the rose family.

Niche "The ecological niche" is for any kind of plant or animal the critical combination of conditions which make it thrive. Examples of its study through field observation and its evolutionary role pp. 145–150.

Pollen mother cells The specialized cells in which the immediate predecessors of male sex cells produce four cells, each with one full set of chromosomes out of their original single cells with two full sets of chromosomes. This is the so-called reduction division.

Polyploid Many-fold, see p. 54.

Recessive See pp. 114–115 for a detailed example of the use of this term from genetics.

Revetment An upright or sloping retaining wall.

Rootstock An underground stem. Though frequently confused with roots, rootstocks are true stems, versatile and sophisticated organs. They contribute greatly to the adaptability of such blessings as cassava (p. 158) and such plagues as poison ivy.

Scatter diagram The oldest (and in critical cases still the most useful) method of studying how two measurable features vary in relation to one another (see Walker, suggested reading).

Stamen The male organ of the higher plants.

Stigma The receptive tissue of the female organ of higher plants.

Tetraploid Four-fold, see p. 54.

Winter cress Barbarea vulgaris Frequently confused with wild mustard but a very different plant. It makes excellent salad greens in late winter and early spring.

Index